GAME THEORY

GAME THEORY

A Nontechnical Introduction

REVISED EDITION

MORTON D. DAVIS

Basic Books, Inc., Publishers   New York

Library of Congress Cataloging in Publication Data

Davis, Morton D., 1930–
   Game theory.

   Bibliography: p. 229
   Includes index.
   1. Game theory.   I.   Title.
   QA269.D38   1983   519.3   83-70771
   ISBN 0-465-02627-3
   ISBN 0-465-02628-1   (pbk.)

To

Gloria, Jeanne, and Joshua

Contents

Contents

Foreword to the First Edition

Oskar Morgenstern

Game theory is a new discipline that has aroused much interest because of its novel mathematical properties and its many applications to social, economic, and political problems. The theory is in a state of active development. It has begun to affect the social sciences over a broad spectrum. The reason that applications are becoming more numerous and are dealing with highly significant problems encountered by social scientists is due to the fact that the mathematical structure of the theory differs profoundly from previous attempts to provide mathematical foundations of social phenomena. These earlier efforts were oriented on the physical sciences and inspired by the tremendous success these have had over the centuries. Yet social phenomena are different:

people are acting sometimes against each other, sometimes cooperatively with each other; they have different degrees of information about each other, their aspirations lead them to conflict or cooperation. Inanimate nature shows none of these traits. Atoms, molecules, stars may coagulate, collide, and explode but they do not fight each other; nor do they collaborate. Consequently, it was dubious that the methods and concepts developed for the physical sciences would succeed in being applied to social problems.

The foundations of game theory were laid by John von Neumann, who in 1928 proved the basic minimax theorem, and with the publication in 1944 of the *Theory of Games and Economic Behavior* the field was established. It was shown that social events can best be described by models taken from suitable games of strategy. These games in turn are amenable to thorough mathematical analysis.

In studying the social world we are in need of rigorous concepts. We must give precision to such terms as utility, information, optimal behavior, strategy, payoff, equilibrium, bargaining, and many more. The theory of games of strategy develops rigorous notions for all of these and thus enables us to examine the bewildering complexity of society in an entirely new light. Without such precise concepts we could never hope to lift the discussion from a purely verbal state and we would forever be restricted to a very limited understanding if, indeed, we could achieve it at all.

It may appear that the mathematical theory remains inaccessible to the mathematically less advanced reader. But this is not so: it is possible to give a clear, comprehensive, and penetrating account of the theory and of many of its applications if an important proviso has been fulfilled. He who attempts to do this, who wishes to give a verbalization of a higher order, must himself have a profound insight into all the intricacies of the theory and, if possible, should have participated in its development. These conditions are amply fulfilled by Morton Davis, the author of this admirable book. A new branch of science is, indeed, fortunate to have found an expositor of his caliber who can bring so many new and in some ways perplexing ideas to the general reader.

A book such as the present is in the best tradition of science writing that has long since been familiar in the natural sciences. In those fields—mathematics not exempted—very able writers have made great efforts to explain accurately but in as simple terms as possible the new results which have come with often bewildering rapidity. In fact, one is probably justified in assuming that these writings have themselves in turn contributed toward further advances since they have spread the knowledge and interest, and many a new mind has been attracted to the respective fields. In the social sciences books such as the present one are rare. This is partly due to the fact that there were few theories to report about comparable in scope and difficulty of those normally encountered in the physical sciences or in that respect comparable to game theory. Partly it was also difficult to avoid the injection of personal value judgments in the discussion of emotionally laden social and economic problems. The present book is entirely free from such faults; it explains, it analyzes, and it offers precepts to those who want to take them; but the theory it describes and develops is neutral on every account.

The reader of this book will be impressed by the immense complexity of the social world and see for himself how complicated ultimately a theory will be that explains it—a theory compared to which even the present-day difficult theories in the physical sciences will pale.

The reader who will follow Davis on his path will be led into a new world of many beautiful vistas; he will see many peaks that have been conquered, but also many still calling for exploration. He will emerge from his trip, I am sure, with a better understanding of the intricacies of our life.

Princeton University
November 1969

Author's Introduction

> We hope, however, to obtain a real understanding of
> the problem of exchange by studying it from an alto-
> gether different angle; this is, from the perspective of
> a "game of strategy."
> —John von Neumann and Oskar Morgenstern
> *The Theory of Games and Economic Behavior*

About forty years ago a mathematician, John von Neumann, and an economist, Oskar Morgenstern, tried to find a more effective way of solving certain kinds of economic problems. They noticed that "the typical problems of economic behavior become strictly identical with the mathematical notions of suitable games of strategy" and so they devised a "Theory of Games." Oddly enough, this new tool turned out to be invaluable in attacking problems in many other areas as well.

In the decades that followed von Neuman and Morgenstern's creation, the mathematical theory has matured and the applications have multiplied—you need only look under the heading "game theory" or "prisoner's dilemma" in the abstracts of psychology, sociology, or political science journals to see the proliferation. In a *Fortune* magazine article John McDonald (1970), writing about executive decision making, noted that game theory is "uniquely qualified to make sense of the forces at work" and described how it could be "related . . . to the strategies of some actual corporations caught up in conglomerate warfare" (p. 122). He saw airline competition, coalition formation to apply political pressure, plant location, product diversification, and conglomerate absorption as fertile areas in which to use game theory.

Elsewhere in the business world game theory is used to derive optimal pricing and competitive bidding strategies and to make investment decisions. It has also been used for picking jurors, measuring a senator's power, committing tanks to battle, allocating business expenses equitably, and as a ploy for animals in an evolutionary struggle.

What, then, is so special about game-theoretical problems? Simply this: in a game there are others present who are making decisions in accordance with their own wishes, and they must be taken into account. While you are trying to figure out what they are doing, they will be trying to figure out what you are doing. If you build a house to be used in all kinds of weather, you have no reason to believe that nature will deliberately make it hot in summer and cold in winter to confound you; if you start a new marketing campaign to increase your sales, however, you can be sure that your competitors will do what they can to frustrate you. In a game each player must assess the extent to which his or her goals match or clash with the goals of others and decide whether to cooperate or compete with all or some of them. It is this blending of players' mutual and conflicting interests that makes game theory fascinating.

This definition of game theory is very broad. There really isn't a "theory" of games; there are in fact many theories.

The nature of the "game," just like the nature of ordinary parlor games, is determined by the "rules." Whether players can communicate and make binding agreements, the information they have, and the ability to share payoffs are all of interest; but more important are the number of players involved and the extent to which their interests coincide or conflict. In fact, these last two criteria determine the structure of the book—so there is a chapter on two-person games with players of conflicting interests, another in which the two players have mixed interests, and yet another on games with more than two players.

In every game we look for a "solution," a description of what each player should do and what the outcome should be. The games in this book get more complex as we proceed, and convincing solutions are harder to come by. It is as though there were a perverse, quasi-conservation law at work: the more important the game, the more applications the game has in the real world, the more difficult it is to analyze. Lying at the "simple" end of the spectrum are games of two players with diametrically opposed interests. These have almost universally accepted solutions. But if, as is generally the case, there are more than two players or the players have both conflicting and joint interests, there may be no solutions, or there may be many. We often settle for outcomes that are more stable, enforceable, or equitable than the others. Although these solutions may be plausible, they are generally not compelling.

Although complex games are less predictable than the simpler ones, they are usually more interesting and more fruitful. By viewing a complex situation as a game, you can translate the intuitive insights of an experienced observer into a quantitative model. This allows you to make subtle quantitative inferences that are far from obvious. It is one thing to say, for example, that a voter's power increases if he or she is given more votes; it is quite another to actually assign a number that reflects this increased power. Not only have various measures of voting power been devised, they have also been applied successfully. The powers of members of congress, the President, and members of various voting bodies such as the United Nations Security Council and the New York City Board of Estimate have all been calculated. And on the basis of just such measures of voting power, a successful challenge of the voting system of Nassau County took place in the courts—the plaintiffs argued that some townships were effectively disenfranchised, and this position was sustained.

Another application of these voting power measures is to so-called bandwagons. The bandwagon phenomenon—the tendency of the preponderance of delegates to flock to the nominee-apparent when his or her success appears imminent—is well known. But by applying these power measures, you can deduce when the voting body will become unstable. This was effectively done in the 1976 Republican primary, and the analysis was more accurate than that of the contemporary observers.

Another, sometimes startling, side benefit of these game-theoretical models is their economy—a model fashioned for one purpose may serve an entirely different one as well. What appear to be two very different kinds of problem turn out to be the same at bottom. Some time ago the following party game was suggested: conduct an auction for something of value—say, a dollar bill—but have both the highest bidder *and the next highest bidder* pay what they bid. It turns out that this is an appropriate model for analyzing an arms race. It also turns out that this model accurately reflects the competition between two male animals competing for a mate. The measure of voting power just discussed is another example of a model serving a dual purpose. The computation that is made to measure a voter's power can be used to equitably apportion costs among airlines that are sharing the same landing strip.

This revised edition differs in a number of ways from the original *Game Theory*, which was published in 1970. A number of problems are posed at the start of each chapter except the first, and readers are given a chance to solve them on their own before reading on. (Unlike most mathematical problems, many problems in game theory are easily understood by the layperson.) At the end of the chapter, where the solutions are discussed, readers can compare their "common sense" solutions with those of the author. Thinking about the solutions to the problems before reading the text should make the chapter all the more rewarding.

In the chapter on two-person, zero-sum games, some useful techniques for finding solutions are included. These techniques are often (but not always) effective and simple, and they frequently yield insights into the nature of the game. A number of applications to real games are mentioned—when best to use a "surprise" serve in tennis and how to win at blackjack—as well as some applications in political science—how to allocate resources in a presidential election and how to pick a jury, for example. Still other applications of mixed strategies are made to the problem of enforcing environmental standards, enforcing laws against speeding, inhibiting crime, and (for animals engaged in evolutionary struggles) maintaining one's territory without becoming engaged in unnecessary conflicts.

Some recent experiments have raised questions about utility theory, the mechanism that allows you to express a person's preferences in a consistent pattern so he or she can make rational decisions. It appears that choices are often made on the basis of the way the alternatives are expressed; while this may not be news to biased poll-takers, salespeople, and advertisers, it *is* disturbing to game theorists. Game theorists assume decision makers behave rationally; when they don't, difficulties arise, and these difficulties are discussed in the text.

Perhaps the most exciting new application of game theory is to biology. Recently considerable work has been done on the role game theory plays in the evolutionary process: studying, for example, how long a fly should wait at one cowpat for a mate before moving on to the next and how aggressive members of a species should be to maximize chances for survival. These and similar questions are discussed at length. The results of two computer tournaments in which the "players" were computer programs written by sophisticated game theorists are described and some of the evolutionary implications are investigated.

The last chapter, on *n*-person games, has been expanded. Two entirely new sections have been added that address the following two questions: How do you set up a voting system that accurately reflects the preferences of the members of society? And once you set up such a system, how does an individual exploit it for his or her own advantage? It turns out that the subject is teeming with pitfalls and paradoxes. A hilarious example of a voting system gone wrong, the Alabama Paradox, is taken from this country's history. A number of new examples applying the theory—the use of Shapley value in accounting procedures, the analysis of bandwagons—are also included.

Despite all the changes, the basic purpose of the book remains the same—to provide a nontechnical explanation of game theory, one of the most fruitful and interesting intellectual products of our time.

Acknowledgments

In the course of writing this book and in the years immediately preceding, I have profited from my contact with many others. From so many others, in fact, that I could not hope to acknowledge them all. Nevertheless, I would like to express my gratitude to those to whom I am particularly indebted. To Professor David Blackwell, my mentor at the University of California, who first introduced me to game theory; to Professor Oskar Morgenstern, who directed the Econometric Research Group at Princeton University where I worked for two years and who allowed me (and many others) access to an intensely stimulating, scholarly environment; to the Carnegie Corporation for financial support during these years; and to the many people with whom I had innumerable conversations—especially Professors Michael Maschler and Robert Aumann of the Hebrew University of Jerusalem.

GAME THEORY

1

An Overview

The theory of games is a theory of decision making. It considers how one should make decisions and, to a lesser extent, how one does make them. You make a number of decisions every day. Some involve deep thought, while others are almost automatic. Your decisions are linked to your goals—if you know the consequences of each of your options, the solution is easy. Decide where you want to be and choose the path that takes you there. When you enter an elevator with a particular floor in mind (your goal), you push the button (one of your choices) that corresponds to your floor. Building a bridge involves more complex decisions but, to a competent engineer, is no different in principle. The engineer calculates the greatest load the bridge is expected to bear and designs a bridge to withstand it.

When chance plays a role, however, decisions are harder to make. A travel agency may want to give its customers prompt service and yet avoid excessive telephone bills. But since the agency doesn't know what future demands will be, it doesn't know how many phones to install. By using past experience and applying the laws of probability, a balance

GAME THEORY

can be struck between losses from excessive phone bills and defecting customers.

Game theory was designed as a decision-making tool to be used in more complex situations, situations in which chance and your choice are not the only factors operating. These are the situations that will concern us from now on. Let's look at a few such examples to clarify what I am saying and postpone our analysis of these kinds of problems to the later chapters.

Companies A and B intend to buy 30 and 24 typewriters, respectively. Salesman P represents the company that currently supplies them both; Q represents a competitor. Each has time to give one sales talk at either company A or B. If they visit the same company, they divide the sales at that company equally but P makes all the sales to the other company. If they visit different companies, each makes all the sales at the company visited. In figure 1.1 the alternatives for salesman Q are represented by columns and the alternatives for P are represented by rows. The entry corresponding to each pair of alternatives represents P's *total sales* (since there are 54 typewriter sales in all, Q's total sales can be deduced from P's sales).

Figure 1.1

		SALESMAN Q	
		VISIT A	VISIT B
SALESMAN P	VISIT A	39	30
	VISIT B	24	42

If Q and P both visit A, for example, we find the entry where column "Visit A" meets the row "Visit A" is 39. P gets half the sales at A and all the sales at B, which is $15 + 24 = 39$. It is understood that Q had 15 sales since the total number of sales is 54.

Generals P and Q each want control of oil deposits that P

now controls. Thirty acres are located at A and 24 are located at B. Q has enough strength to invade only one location and P has enough strength to defend only one. Both forces are equally matched, so if P and Q go to the same place there will be a standoff; each gets half the acreage there while P retains control of the other location. If P and Q choose different locations, each army will control all the acreage at its own location.

Figure 1.2 indicates the total acreage that P will control.

Figure 1.2

		GENERAL Q	
		A	B
GENERAL P	A	39	30
	B	24	42

On the last day of a political convention, aspiring candidates P and Q will meet delegates from either states A or B. P is the current favorite, so if both candidates visit the same state, each will get half the delegates of that state and P will get all the delegates of the other one; if they visit different states, each will get the delegates of the state visited. If A and B have 30 and 24 delegates, respectively, the possible outcomes are reflected in figure 1.3.

Figure 1.3

		ASPIRING CANDIDATE Q	
		A	B
ASPIRING CANDIDATE P	A	39	30
	B	24	42

Although these three situations differ—one involves business competition, another military conflict, and the third a political campaign—they all reduce to a single problem involving the theory of games. They differ from the problems described earlier—building a bridge and installing tele-phones—in one essential respect: *While decision makers are trying to manipulate their environment, their environment is trying to manipulate them.* A store owner who lowers her price to gain a larger share of the market must know that her competitors will react in kind. A thief who robs banks rather than newsstands (because that's where the money is) must be aware that the police will be asking themselves "Where would I go if I were a thief?" and act accordingly. The bridge under construction, on the other hand, has no feelings about its own safety, and the travel agency's customers were not trying to embarrass it by calling too frequently or not calling it frequently enough.

A player involved in a game with other decision-making players is in the same position as the scientist who wanted to study a monkey's behavior. After he placed the monkey in a room and gave it time to get acclimated, he looked through the observation slot—and saw the monkey's eyeball looking back at him.

In the rest of this book I will talk about certain situations that I will call *games.* In a game there will be *players* (at least two), and each will pick a *strategy* (make a decision). As a result of this joint choice—and possibly chance, a disinterested player—the result will be a reward or punishment for each player: the *payoff.* Because everyone's strategy affects the outcome, a player must worry about what everyone else does and knows that everyone else is worrying about him or her.

The words "strategy," "player," and "payoff" have rough-ly the same meaning here as they do in everyday language. A *player,* a participant in the game, need not be a single person. If each member of a group has exactly the same feelings about how the game should turn out, the members may be considered a single player. So a "player" may be a corpora-tion, a county, or a football team.

A *strategy* in game theory is a complete plan of action that describes what a player will do under all possible circumstances. In ordinary usage a strategy is considered to be something clever, but nothing like that is intended here. There are poor strategies, just as there are good strategies. In the three examples depicted in figures 1.1 to 1.3, each player had two simple strategies—*A* and *B*—but in real games strategies may be so complex that they can't be written out explicitly. Also, in some real games it might seem convenient to think of a player as using several different strategies. Competing automobile companies that fix their prices every year and chess players who rethink their position after each move are two examples. But in principle you can imagine that all these decisions are merged into one to form a single strategy, and from the point of view of the theorist this is more convenient.

A complete strategy in chess, then, would start something like this: "On my first move I will move to position *A*. If he then moves to *B*, then I'll move to *B'*; if he then moves to *C*, I'll move to *C'*; if he . . . If after I move to *A*, he moves to *B*, I move to *B'*. If he then moves to *Q*, I'll move to *Q'* . . ." It is almost impossible to describe a complete strategy in detail in any real game one actually plays, and even in such a simple game as tick-tack-toe, which is played by very young children, the task is formidable. But the practical problem of writing out an entire strategy in detail shouldn't stop us from making use of the concept any more than our inability to multiply all the numbers from 1 to 1,000 shouldn't stop us from writing a formula in which this product appears.

This distinction between theory and practice is very important; the difference becomes clear if you compare a chess player's attitude toward the game of chess with that of the game theorist. Chess has been around for many centuries, and no human being or computer has come close to mastering it. In films and cartoons the bearded chess player is often used to symbolize profound thought, and, in fact, the chess player finds the game profound and subtle. To the game theorist, however, the game is trivial. This seems an absurd position, since game theorists are not even particularly good chess players.

The apparent paradox is easily resolved, however; chess, complex as it is, is finite, so, *in principle,* every position on the board is either (a) a win for white, (b) a draw, or (c) a win for black. Given enough time, you could start from the end of the game and work forward labeling every position that can possibly arise in the game as a win, loss, or draw and finally determine whether the game itself is a win, loss, or draw. (This technique would also indicate how to enforce the win or draw.) In practice the task is hopeless, of course, so from the chess player's point of view the game is as deep as he or she believes it to be.

There is also a difference in the way we will be analyzing our "games" and the way experts analyze parlor games. If a player in a game like chess has a winning strategy, we assume he will use it; if a player is in great difficulty and needs to adopt a very subtle defense to draw, we assume she will find it. In short, we assume players always do their best.

In real life, it may make an enormous difference how you play even if you are backing a lost cause. Against an ideal player you may be defeated, but against a real person it is well known that certain strategies induce errors. Emanuel Lasker, world chess champion for many years, felt that psychology plays a very important role in the game. He often adopted a slightly inferior opening, which initially gave him a slight disadvantage, in order to disconcert his opponent. And a Russian handbook on championship chess suggests that a player should try to force an opponent into an early commitment, even if by doing so the first player obtains a slightly inferior position. In the children's game of tick-tack-toe, the outcome will always be a draw if there is correct play on both sides, but there is a pragmatic argument for making the first move in the corner: there is only one answer to the corner move that preserves the draw, and that is a move in the center. Any other first move allows at least four adequate replies. So, in a sense, the corner move is strongest, but it is a sense that the game theorist does not recognize. Game theorists do not speak of "slightly unfavorable positions" or "commitments" or "attacks," premature or otherwise. They are incompetent to deal with the game on these terms, and these terms are superfluous to their theory. In short, game theorists, do not attempt to exploit their opponent's folly.

Since it takes no great insight to recognize the existence of folly in this world, and since the game theorist purports to be influenced by the world, why this puristic attitude? The answer is simply this: it is much easier to recognize the existence of error than to fashion a general, systematic theory that will allow you to exploit it. So the study of tricks is left to the experts in each particular game; game theorists make the pessimistic, and often imperfect, assumption that their opponents will play flawlessly.

Games like chess, checkers, tick-tack-toe, and the Japanese game of Go are called *games of perfect information* because everyone knows exactly what is going on at all times. These games offer few conceptual problems, and they won't be discussed here. In games like poker and bridge the players are, to some extent, kept in the dark, and in this sense the games are more complex. Even as trivial a game as matching pennies in which each player must choose a strategy without knowing what an adversary is doing has this added dimension of complexity.

One of the engaging properties of game theory is that many problems can be understood immediately without any special technical background. Terms such as "zero-sum" and "prisoner's dilemma" have become part of the vocabulary of economics and the social sciences. Particularly seductive are the problems growing out of chapters 5 and 6, both for the sophisticate and beginner. Since much of the material is accessible to the reader, a number of problems have been listed at the beginning of each chapter instead of the end. Read them and think about them before your ideas are shaped by the text. They generally don't require any great quantitative skill, but they do require thought. This forethought will make the book much more challenging.

2

The Two-Person, Zero-Sum Game with Equilibrium Points

Introductory Problems

Several variations of a certain type of game are shown in figures 2.1 to 2.4. Take a moment to see what you do in each case and guess what the likely outcome will be. The material in the chapter will be much more interesting if you think about the ideas involved first.

Each of the matrices shown below represents a game. I will describe how game 1 is played in some detail; the other games are played in much the same way.

You pick a row (*A*, *B*, or *C*) and your opponent picks a column (*I*, *II*, or *III*) at the same time so neither knows when choosing what the other has picked. The number where the row and column intersect is the amount your opponent *pays*

Figure 2.1

YOUR OUTCOME

		I	II	III
	A	5	−2	1
YOU	B	6	4	2
	C	0	7	−1

Figure 2.2

YOUR OUTCOME

		I	II	III
	A	−2	1	1
YOU	B	−3	0	2
	C	−4	−6	4

Figure 2.3

YOUR OPPONENT

		I	II	III	IV
	A	−3	17	−5	21
YOU	B	7	9	5	7
	C	3	−7	1	13
	D	1	−19	3	11

Figure 2.4

YOUR OPPONENT

		I	II	III
	A	2	−5	−2
YOU	B	3	−1	−1
	C	−3	4	−4

you in dollars. So if you pick row *A* and your opponent picks column *III*, you will receive a dollar from her. (If your opponent chose *II* you would pay two dollars to her, since the number is negative.) You may assume here and throughout the book that your opponent knows the rules of the game and is as intelligent as you are. Remember that you must take your opponent's thinking into account—if you play *C* you have a chance at your greatest possible gain, 7, but will your opponent cooperate by choosing column *II*? So once again: What do you do, why, and what should the outcome of this game be?

In each of the first four games, imagine what you would do if you *knew* in advance your opponent's strategy (take each of his or her strategies in turn). If your choice depends on your opponent's choice, how do you play when you don't know what he or she will do?

Figure 2.5 depicts a similar type of game but some of the payoffs (matrix entries) have been omitted. Can you still predict what will happen even though you don't know the missing payoffs?

Figure 2.5

YOUR OPPONENT

		I	II	III
YOU	A	?	?	3
	B	?	?	4
	C	7	6	5

IN FEBRUARY 1943 General George Churchill Kenney, Commander of the Allied Air Forces in the Southwest Pacific, was faced with a problem. The Japanese were about to reinforce their army in New Guinea and had a choice of two alternative routes. They could sail either north of New Britain, where the weather was rainy, or south of New Britain, where the weather was generally fair. In any case, the journey would take three days. General Kenney had to decide where to concentrate the bulk of his reconnaissance aircraft. The Japanese wanted their ships to have the least possible exposure to enemy bombers, and, of course, General Kenney wanted the reverse. The matrix entries in figure 2.6 represent the expected number of days of bombing exposure.

It is more difficult to make decisions in this kind of a game than in the games mentioned in chapter 1. The critical difference between this game and the game of chess is that here the players lack information. Both players must decide simul-

Figure 2.6

		JAPANESE CHOICE	
		NORTHERN ROUTE	SOUTHERN ROUTE
ALLIED CHOICE	NORTHERN ROUTE	2 days	2 days
	SOUTHERN ROUTE	1 day	3 days

taneously, so neither knows the other's strategy when choosing his or her own. The analysis of this particular game is simple enough, however. At first it seems the Allies have a problem: it would be best for them to take the same route as the Japanese, but when they make their decision they don't know what that route will be. But the problem is quickly solved when you take the Japanese view. For them, the northern route minimizes their exposure *whatever* the Allies do, so their action is clear. After working this out, the Allies' decision also becomes clear: go north.

This last is an example of a two-person, zero-sum game with equilibrium points. The term "zero-sum" (or equivalently, "constant sum") means the players have diametrically opposed interests. The term comes from parlor games like poker where there is a fixed amount of money around the table. If you want to win some money, others have to lose an equivalent amount. Two nations trading make up a non-zero-sum game since both may simultaneously gain. An equilibrium point is a stable outcome of a game associated with a pair of strategies. It is considered stable because a player unilaterally picking a new strategy is hurt by the change.

A Political Example

It is an election year and the two major political parties are in the process of writing their platforms. There is a dispute between state X and state Y concerning certain water rights. Each party must decide whether it will favor X or favor Y or evade the issue. The parties, after holding their conventions privately, will announce their decisions simultaneously.

Citizens outside the two states are indifferent to the issue.

In X and Y, the voting behavior of the electorate can be predicted from past experience. The regulars will support their party in any case. Others will vote for the party supporting their state or, if both parties take the same position on the issue, will simply abstain. The leaders of both parties calculate what will happen in each circumstance and come up with the matrix shown in figure 2.7. The entries in the matrix are the percentage of votes party A will get if each party follows the indicated strategy. If A favors X and B dodges the issue, A will get 40 percent of the vote.

Figure 2.7

		B'S PLATFORM		
		FAVOR X	FAVOR Y	DODGE ISSUE
	FAVOR X	45%	50%	40%
A'S PLATFORM	FAVOR Y	60%	55%	50%
	DODGE ISSUE	45%	55%	40%

This is the simplest example of this type of game. Though both parties have a hand in determining how the electorate will vote, there is no point in one party trying to anticipate what the other will do. Whatever A does, B does best to dodge the issue; whatever B does, A does best to support Y. The predictable outcome is an even split. If, for some reason, one of the parties deviated from the indicated strategy, this should have no effect on the other party's actions. A slightly more complicated situation arises if the percentages are changed a little, as shown in figure 2.8.

B's decision is now a bit harder. If it thinks A will favor Y, it should dodge the issue; otherwise, it should favor Y. But the answer to the problem is in fact not far off. A's decision is clear-cut and easy for B to read: favor Y. Unless party A is

GAME THEORY

Figure 2.8

		B'S PLATFORM		
		FAVOR X	FAVOR Y	DODGE ISSUE
	FAVOR X	45%	10%	40%
A'S PLATFORM	FAVOR Y	60%	55%	50%
	DODGE ISSUE	45%	10%	40%

foolish, B should realize that the chance of getting 90 percent of the vote is very slim—indeed, not a real possibility—and that it would do best to dodge the issue.

This is the same type of situation that General Kenney had to contend with. On the face of it, the northern and southern routes both seemed plausible strategies. But the rainy northern route was obviously more favorable for the Japanese, which meant that the northern route was the only reasonable strategy for the Allies.

In figure 2.9, neither player has an obviously superior strategy. In this case, both players must think a little. Each player's decision hangs on what he or she expects the other will do. If B dodges the issue, A should too. If not, A should favor Y. On the other hand, if A favors Y, B should favor X. Otherwise, B should favor Y.

Once again, the underlying structure is not hard to analyze. While at first B may not be clear about what should be done, it's obvious what B should *not* do: B should not dodge

Figure 2.9

		B'S PLATFORM		
		FAVOR X	FAVOR Y	DODGE ISSUE
	FAVOR X	35%	10%	60%
A'S PLATFORM	FAVOR Y	45%	55%	50%
	DODGE ISSUE	40%	10%	65%

16

the issue, since whatever A does, B always does better if he or she favors X over dodging the issue. Once this is established, it immediately follows that A should favor Y and, finally, that B should favor X. A will presumably wind up with 45 percent of the vote.

These two strategies—A's favoring Y and B's favoring X—are important enough to be given a name: *equilibrium strategies.* The outcome resulting from the use of these two strategies—the 45 percent vote for A—is called an *equilibrium point.*

What are equilibrium strategies and what are equilibrium points? Two strategies are said to be in equilibrium (they come in pairs, one for each player) if neither player gains by changing strategy unilaterally. The outcome (sometimes called the payoff) corresponding to this pair of strategies is defined as the equilibrium point. As the name suggests, equilibrium points are very stable. In two-person, zero-sum games, at any rate, once the players settle on an equilibrium point, they have no reason for leaving it. If, in the last example, A knew in advance that B would favor X, A would still favor Y; and, similarly, B would not change strategy if he or she knew A would favor Y. There may be more than one equilibrium point, but if there is, they will all have the same payoff.

This is true in two-person, zero-sum games, that is. Equilibrium strategies and equilibrium points can be spoken of in n-person and non-zero-sum games, but the arguments favoring them are considerably less appealing. In a two-person, non-zero-sum game, for example, equilibrium points need not have the same payoff; one equilibrium point may be more attractive to *both* players than another. This will be discussed in more detail in the section on the prisoner's dilemma, in chapter 5.

If they exist, equilibrium points are easy to find. In the

game shown in figure 2.9, assume that *B* knows *A*'s strategy in advance. Since *B* would choose the *minimum* value of any row *A* chooses, *A* should choose a strategy that yields the *maximum* of these *minimum* values; this value is called the *maximin,* and it is the very least that *A* can be sure of getting. In this game, if *A* plays "Favor *X*," "Favor *Y*," and "Dodge Issue," these minimum values would be 10, 45 and 10, respectively, and the maximin would be 45.

Now imagine the rules are changed so that *A* knows *B*'s strategy in advance. *A* would be expected to choose the maximum of any column, so *B* should choose the column that minimizes these maxima. That outcome is called the *minimax;* in this game the maximum values associated with the "Favor *X*," "Favor *Y*," and "Dodge Issue" strategies of *B* are 45, 55, and 65, respectively, so the minimax is 45.

If the minimax equals the maximin, the payoff is an equilibrium point and the corresponding strategies are an equilibrium strategy pair.

An equilibrium point and its corresponding equilibrium strategies are easily recognized once they are pointed out. The payoff associated with an equilibrium point is the *smallest in its row* and *the largest in its column.* In figure 2.10 (a

Figure 2.10

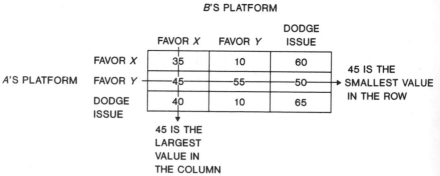

repetition of figure 2.9), the equilibrium payoff of 45 is smallest in its row and largest in its column.

When an equilibrium point exists in a two-person, zero-sum game, it is called the *solution*. Rational players should adopt the equilibrium strategies and the outcome should be the payoff associated with the equilibrium point—the *value* of the game. In the game just discussed, the equilibrium strategies for players A and B were "Favor Y" and "Favor X," respectively, and the value of the game was 45. The reasons why equilibrium points are considered solutions are:

1. *By playing his or her equilibrium strategy, a player will get at least the value of the game.* In the game shown in figure 2.10, A gets at least 45, whatever B does, if A plays "Favor Y."
2. *By playing his or her equilibrium strategy, an opponent can stop a player from getting any more than the value of the game.* By playing "Favor X," B can limit A's payoff to 45 whatever A does.
3. *Since the game is zero-sum, a player's opponent is motivated to minimize the player's payoff.* When A gets 45, B gets 55; if A gets any more, it must be because B obtained that much less.

In games with equilibrium points, payoffs that are not associated with either equilibrium strategy have no bearing on the outcome. If in figure 2.10 the two payoffs of 10 and the payoffs of 60 and 65 were changed in *any way whatever*, the players should pick the old equilibrium strategies and the outcome should be the same.

GAME THEORY

Domination

It is often possible to simplify a game by eliminating *dominated* strategies.

Strategy *A* dominates strategy *B* if a player's payoff with strategy *A* is (a) always at least as much as that of strategy *B* (whatever other players do) and, (b) at least some of the time actually better than strategy *B*. Consider the game in figure 2.11, for example:

Figure 2.11

		YOUR OPPONENT		
		I	*II*	*III*
	A	7	9	8
YOU	B	9	10	12
	C	8	8	8

For you, strategy *B* dominates both strategies *A* and *C* because it always yields a higher payoff. Your opponent's strategy *I* dominates strategies *II* and *III* (recall that the matrix entries represent what your opponent pays you, so he or she wants them *small*). Although your opponent doesn't always do better with *I* than with *II* or *III*, he or she always does at least as well, and sometimes does better.

Whenever you analyze a zero-sum game, you may assume:

1. *You will never pick a dominated strategy*—why pick a dominated strategy when you can do at least as well using the strategy that dominates it?
2. *Your opponent will never pick a dominated strategy*—and this for the same reason that you won't.

In fact, if your opponent plays a dominated strategy, you can only be pleasantly surprised—you will do as well as you would have had your opponent picked the dominating strategy, and you were already resigned to that.

If all strategies but one are dominated for each player, the equilibrium point(s) can be calculated. Consider the game shown in figure 2.12, for example:

Figure 2.12

		YOUR OPPONENT		
		I	*II*	*III*
	A	10	0	1
YOU	*B*	11	9	3
	C	23	7	−3

In this game none of your strategies are dominated initially; but since *III* dominates *I* for your opponent you can eliminate *I* from consideration. With *I* eliminated, *B* dominates *A* and *C*, and with *A* and *C* eliminated, *III* dominates *II*. The only undominated strategies, *B* and *III*, make up an equilibrium strategy pair and the value of the game is 3.

If we go back to our earlier examples, we find a number of dominated strategies. In figure 2.6 the northern route dominated the southern route for the Japanese. After eliminating the Japanese southern route, we eliminated the Allied southern route for the same reason. In figure 2.7 "Favor *Y*" dominated all of *A*'s other strategies and "Dodge Issue" eliminated all of *B*'s. In figure 2.8 "Favor *Y*" dominated everything for *A* and then "Dodge Issue" dominated everything for *B*. And finally, in figure 2.9, "Favor *X*" dominated "Dodge Issue" for *B*, "Favor *Y*" dominated everything for *A*, and "Favor *X*" dominated "Favor *Y*" for *B*.

If a game has an equilibrium point, it is easy to choose appropriate strategies and predict the outcome. But what if there is no equilibrium point? Take the simple game of matching pennies, for example as depicted in figure 2.13.

Figure 2.13

		OPPONENT	
		HEADS	TAILS
YOU	HEADS	−1	+1
	TAILS	+1	−1

Since no strategy is dominated and there is no equilibrium point, it is hard to see how you can play such a game rationally. To try to construct a theory for such a game would appear to be a waste of time. Von Neumann and Morgenstern (1953), put the problem this way:

Let us imagine that there exists a complete theory of the zero-sum two-person game which tells a player what to do and which is absolutely convincing. If the players knew such a theory then each player would have to assume that his strategy has been "found out" by his opponent. The opponent knows the theory, and he knows that a player would be unwise not to follow it. Thus the hypothesis of the existence of a satisfactory theory legitimizes our investigation of the situation when a player's strategy is "found out" by his opponent. (p. 148)

The paradox is this: if we are successful in constructing a theory that indicates which strategy is best, an intelligent opponent with access to all the information available to us can use the same logic to deduce our strategy. The opponent can then second-guess us and win. So it would be fatal to use the "preferred" strategy suggested by the theory.

In fact, however, you can construct a theory that enables you to play such games intelligently. Such a theory will be the focus of the next chapter.

Solutions to Problems

1. Your opponent should see that your *B* strategy dominates your *A*—that is, you do better with *B* *whatever* your opponent does. Once your opponent rules out *A* as a possible strategy, he or she does best with *III*. You, in turn, do best with *B*. *III* and *B* are the suggested strategies, and you should receive a payoff of 2.

2. *I* and *A* are the suggested strategies, and you should pay 2.

3. *B* and *III* are the suggested strategies, and you should be paid 5.

4. *B* and *III* are the suggested strategies, and you should pay 1.
   In each of games 1 through 4, both players will *not* lose if they announce their strategy in advance to their opponent.

5. Whatever the values of the missing payoffs, you can be sure of getting 5 by playing *C* and your opponent can be sure of losing no more than 5 by playing *III*. Since either one of you can enforce the payoff of 5, this is a plausible outcome of the game no matter what the missing payoff entries are.

3

The General, Two-Person, Zero-Sum Game

Introductory Problems

In the last chapter several games were described in which the fate of each player was, at least to some extent, in the hands of the opponent. But in each case there was an equilibrium point so that each player, by his or her own efforts, could get the value of the game, which was all that could reasonably be expected. The games in this chapter have no equilibrium points; if you are to win anything like the amount that you should, you must start second-guessing your opponent in earnest.

1. Real variations of poker are too complex to analyze so in actual games decisions are made by feel and experience. In

the simplified poker game described in figure 3.1, decide what strategy you and your opponent should choose (based on your intuition) and estimate the outcome.

You and your opponent each toss a coin with a "1" on one side and a "0" on the other. After examining your own coin, you may either *bid* or *pass*. If you pass, the player with the higher number wins $2 from the other—no money passes if there is a tie. If you bid, your opponent can *see* or *fold*. If your opponent folds, you win $11; if your opponent sees, the higher number wins $12 (again, ties are a standoff).

Figure 3.1

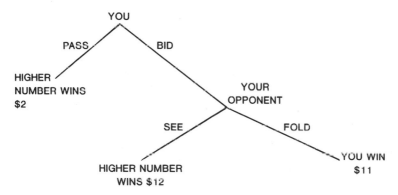

2. Two suspected smugglers are about to flee from either the sole airport or the sole seaport of a country; two police officers are assigned to catch them. If both officers guard one exit and the smugglers use the other, 100 pounds of contra-

Figure 3.2

		NUMBER OF POLICE OFFICERS AT THE AIRPORT		
		0	1	2
NUMBER OF SMUGGLERS AT THE AIRPORT	0	0	70	100
	1	50	0	50
	2	100	70	0

band will get through; if one officer guards the exit used by both smugglers, 70 pounds will get through; and if there is one smuggler at a port with no police officer, 50 pounds will get through. If there are at least as many police officers as smugglers at a port, nothing will get through. Assuming the goal of the police officers and the smugglers is to minimize and maximize the smuggled contraband, respectively, what should each do and how much will be smuggled?

3. A company's cash is contained in two safes, which are kept some distance apart. There is $90,000 in one safe and $10,000 in the other. A burglar plans to break into one safe and have an accomplice set off the alarm in the other one. The watchman has time to check only one safe; if he guards the wrong one, the company loses the contents of the other safe, and if he guards the right one, the burglar leaves empty-handed. From which safe is a sophisticated burglar more likely to steal? With what probability? What should the watchman do, and how much on average, will be stolen?

4. Your team is ahead in a team-of-four bridge tournament with one deal left. If you and your opponents in the other room reach the same contract, you win the tournament. If one team bids the unlikely slam while the other bids game, the slam bidder has a 10 percent chance of winning the tournament. What are the appropriate strategies for you and your opponents and what should the outcome be? (Note: matrix entries reflect your chance of winning the tournament.)

Figure 3.3

		YOUR OPPONENTS' BID	
		SLAM	GAME
YOUR BID	SLAM	1	.1
	GAME	.9	1

THE two-person games discussed in the last chapter were easily analyzed. They all had equilibrium points, so each player could choose a strategy that guaranteed him or her the value of the game. In this, the chapter was misleading; all but the simplest zero-sum games that arise in practice do not have equilibrium points, and they present new and formidable problems. Edgar Allan Poe (1902) discussed such a game—an extremely simple one—in *The Purloined Letter*, and it is interesting to compare his approach with von Neumann's, which will be discussed later. Here is Poe:

"I knew one [student] about eight years of age, whose success at guessing in the game of 'even and odd' attracted universal admiration. This game is simple, and is played with marbles. One player holds in his hand a number of these toys, and demands of another whether that number is even or odd. If the guess is right, the guesser wins one; if wrong, he loses one. The boy to whom I allude won all the marbles of the school. Of course he had some principle of guessing; and this lay in mere observation and admeasurement of the astuteness of his opponents. For example, an arrant simpleton is his opponent, and, holding up his closed hand, asks, 'are they even or odd?' Our schoolboy replies, 'odd,' and loses; but upon the second trial he wins, for he then says to himself, 'the simpleton had them even upon the first trial, and his amount of cunning is just sufficient to make him have them odd upon the second; I will therefore guess odd;'—he guesses odd, and wins. Now, with a simpleton a degree above the first, he would have reasoned thus: 'This fellow finds that in the first instance I guessed odd, and, in the second, he will propose to himself, upon the first impulse, a simple variation from even to odd, as did the first simpleton; but then a second thought will suggest that this is too simple a variation, and finally he will decide upon putting it even as before. I will therefore guess even;'—he guesses even, and wins. Now this mode of reasoning in the schoolboy, whom his fellows termed 'lucky,'—what, in its last analysis, is it?"

"It is merely," I said, "an identification of the reasoner's intellect with that of his opponent."

"It is," said Dupin; "and, upon inquiring of the boy by what means he effected the *thorough* identification in which his success consisted, I received answer as follows: 'When I wish to find out how wise, or how stupid, or how good, or how wicked is any one, or what are his thoughts at the moment, I fashion the expression of my face, as accurately as possible, in accordance with the expression of his, and then wait to see what thoughts or sentiments arise in my mind or heart, as if to match or correspond with the expression.'" (p. 123)

"As poet *and* mathematician, he would reason well; as mere mathematician, he could not have reasoned at all, and thus would have been at the mercy of the Prefect." (p. 127)

It will be easier to appreciate von Neumann's approach to this problem if we look a little more at the problem first. Put yourself in the shoes of one of the schoolboys who was pitted against Poe's remarkable wonder child.

Your situation seems hopeless; every idea that occurs to you will occur to the wonder child too. You can try to fool him by fixing your face into an "even configuration" and then playing odd. But how do you know that what you think of as an "even" configuration is not really the expression of someone playing odd and trying to look as though he were playing even? Perhaps you ought to look even, think even, and then (slyly) *choose* even. But once again, if you are clever enough to figure out this tortuous plan, won't he be clever enough to see through it? This kind of reasoning can be extended to any number of steps and the effort would be as futile as it is exhausting.

Let's look at the payoff matrix of this game in figure 3.4. The numbers in the matrix indicate the number of marbles that you are paid by the wonder child (or that you pay, if the number is negative).

Now, instead of trying to figure out just how much insight the wonder child has, suppose you assume the worst—he is so clever that he can anticipate your thinking on every count. In effect, you must announce your strategy in advance, and the wonder child can use the information as he wishes. In this case, it seems to make little difference what you do. Whether you choose even or odd, the result will be the same: the loss of a marble. Thus, if the wonder child has perfect insight, it appears that you may as well resign yourself to the loss of a marble. You certainly can't do any worse than this—but is there a chance of doing better?

Figure 3 4

		POE'S STUDENT	
		EVEN	ODD
YOU	EVEN	−1	+1
	ODD	+1	−1

The fact is that you *can* do better, despite your opponent's cleverness. And, ironically, the means of doing so is inadvertently suggested by the second quotation from Poe: *the way to do better is not to reason at all.* To see why, let's go back a bit.

I said earlier that, as chooser, you have only these alternatives: to pick an even number of marbles or an odd number of marbles. In one sense this is true; ultimately you have to make one of these two choices. But in another sense this is *not* true. There are many *ways* in which you can make this choice. And although it may seem that *what* you decide is all that matters and *how* you decide is irrelevant, how you decide is actually critical. Of course, you can always pick one of the *pure* strategies: odd or even. But you can also use a chance device such as a die or roulette wheel to make the decision for you. Specifically, you might throw a die and if a six appears choose even and otherwise choose odd. A strategy that prescribes the selection of a pure strategy by means of a random device is called a *mixed* strategy.

From the second point of view, you have not two but an

29

infinite number of mixed strategies: you can choose an odd number of marbles with probability p and an even number of marbles with probability $1-p$ (p being a number between zero and 1). Roughly speaking, p is the fraction of times an odd number would be chosen if the game were played an infinite number of times. The pure strategies—"always play even," "always play odd"—are the extreme cases, where p is zero and 1, respectively.

Returning to *The Purloined Letter,* let us suppose you play even half the time; you might flip a coin and play odd whenever it comes up heads. Suppose the wonder child guesses that p is one-half, or you tell him. There is nothing more he can learn from you—you simply don't know any more. Also, unless he has powers undreamed of even by Poe, he can't possibly predict the outcome of a random toss of a coin. If you pick this mixed strategy, the outcome will be the same no matter what your opponent does; that is, each of you will win, on the average, half the time.

To summarize: you have taken a game in which you were seemingly at the mercy of a clever opponent and transformed it into one in which your opponent had essentially no control over the outcome. But if you can do as well as this when you give away your strategy, wouldn't you do better if you kept it to yourself? No, you can't do any better, and it's clear why: your opponent can do the same thing you have done. By using a random device, he too can ensure an even chance of winning. So *each* player has it within his power to see that no advantage is gained by the other.

It is conceivable that a player may do better by not "randomizing." In the game described by Poe, for example, if a player has a talent for anticipating his opponent's choices, he might try to exploit that talent. If your opponent doesn't know about randomizing (or, mistakenly, feels he is cleverer than you), it may be possible to exploit his ignorance. But this is a doubtful advantage at best: it can always be neutralized by a knowledgeable opponent.

It should also be noted that in the odd-even game, once a player resorts to randomizing, not only can he or she not lose (on average), no matter how well an opponent plays; he or she will not gain, *however badly an opponent plays*. The more capable your opponent, then, the more attractive the randomizing procedure.

The reasoning on which the analysis of the odd-even game is based is applicable in much more complicated situations as well.

A Military Application

General X intends to attack one or both of two enemy positions, A and B; and General Y must defend them. General X has five divisions at his disposal; General Y has three. Each general must divide his entire force into two parts and commit them to these two positions, without knowing what the enemy will do. The outcome is determined once the strategies are chosen. The general who commits the most divisions to any one position will win there; if both generals send the same number of divisions, each will get credit for half a victory. Assuming that each general wishes to maximize his expected number of victories (that is, the average), what should he do? What is the expected outcome?

The entries in the matrix in figure 3.5 represent the average number of victories won by General Y. If General Y sends two divisions to A, for example, and General X does the same, General Y will gain half a victory at A and no victory at B, where he is outnumbered three to one.

Let's start by looking at the problem from General Y's

point of view. Assume he's a pessimist and he believes General X is capable of second-guessing him. It is clear that whatever strategy he picks, he will have no victories; General X will assign one more division to A than General Y did and have one extra division at B as well; that is, the maximin is 0.

Figure 3.5

			GENERAL X'S STRATEGIES DIVISIONS SENT TO A					
			0	1	2	3	4	5
		0	½	0	½	1	1	1
GENERAL Y'S	DIVISIONS	1	1	½	0	½	1	1
STRATEGIES	SENT TO A	2	1	1	½	0	½	1
		3	1	1	1	½	0	½

Now suppose General Y decides to play a mixed strategy; that is, he makes his decision on the basis of a chance device. We will still assume that General X can guess the nature of the chance device (the probabilities of selecting each of the strategies), but not the actual outcome of the toss of the coin or spin of the wheel. Specifically, suppose General Y assigns all his divisions to A one-third of the time, all his divisions to B one-third of the time, and makes each of the two possible two/one splits one-sixth of the time.* General X, aware of General Y's strategy, need examine only what will happen in each of his own six strategies and then select the one that gives him the greatest advantage. It is a matter of simple calculation to determine that General X will obtain an average of 1 5/12 victories (and, correspondingly, General Y will obtain an average of 7/12 victories) if he does not put all his

---

*I am primarily concerned with describing the meaning and significance of the von Neumann solution. (The technique of computing the strategies and value of the game will be discussed later.) For now, I will pull solutions out of a hat without apology whenever appropriate.

divisions in one location, and he will obtain an average of 1 1/6 victories otherwise.

What can we conclude from all this? Simply that General Y can reasonably expect to win an average of 7/12 victories, no matter how clever General X is. In fact, he can win the 7/12 victories even if he tells General X his strategy in advance. But should he settle for this? To answer this question, let us look at the problem from General X's point of view.

It is clear from the start that if General X commits himself to a pure strategy and General Y figures it out, they will each get one victory. When five divisions are distributed between two locations, there must be two or less divisions at one of the locations. If General Y sends all three of his divisions to the location where General X sent his smaller force, he will gain precisely one victory (and there is no way he can do any better); that is, the minimax is one.

But now suppose General X sends one, two, three, four divisions to A, with probabilities one-third, one-sixth, one-sixth, one-third, respectively, and the balance of his forces to B. (A little thought should make it obvious that it never pays to send all five divisions to one location.) It makes no difference what General Y does. The result will be the same: General Y will obtain an average of 7/12 victories.

In short, the situation is this. Either General X or General Y can, without the aid of, and without errors from, his opponent, ensure an outcome at least as favorable to himself as the outcome just mentioned: 7/12 of a victory (on average) for General Y, 1 5/12 victories for General X. The answer to the earlier question is that neither general can hope to do better than this against an informed opponent.

A Marketing Example

Two firms are about to market competitive products. Firm *C* has enough money in its advertising budget to buy two blocks of television time, an hour long each, while firm *D* has enough money for three such one-hour blocks. The day is divided into three basic periods: morning, afternoon, and evening, which I will indicate by *M, A,* and *E,* respectively. Purchases of time must be made in advance and are kept confidential. Fifty percent of the television audience watch in the evening; 30 percent watch in the afternoon; and the remaining 20 percent watch in the morning. (For the sake of simplicity, it is assumed that no person watches during more than one period.)

If a firm buys more time during any period than its competitor, it captures the *entire* audience during that period. If both firms buy the same number of hours during any one period—and this is so if neither firm buys time at all—each gets half the audience. Each member of the audience buys the product of just one of the firms. How should the firms allocate their time? What part of the market should each firm expect to get?

The payoff matrix for this game is shown in figure 3.6. The strategies for firm *C* (there are six) are indicated by two letters. *MA* means one hour of advertising in the morning and another hour in the afternoon. Similarly, the ten strategies for firm *D* are expressed by three letters. The entries in the matrix represent the percentage of the market that firm *D* captures; firm *C* gets the rest.

To illustrate how the percentage entries were obtained, assume firm *D* chooses *EMM*—one evening and two morning hours—and firm *C* chooses *EM*—one evening and one morning hour. Since firm *D* bought two morning hours to firm *C*'s one, firm *D* will capture all of the morning market of

Figure 3.6

C'S STRATEGIES

	EE	EA	EM	AA	AM	MM
EEE	75	60	65	60	50	65
EEA	65	75	80	60	65	80
EEM	60	70	75	70	60	65
EAA	40	65	55	75	80	80
EAM	50	60	65	70	75	00
EMM	35	45	60	70	70	75
AAA	40	40	30	65	55	55
AAM	50	50	40	60	65	55
AMM	50	35	50	45	60	65
MMM	35	20	35	45	45	60

D'S STRATEGIES

20 percent. Since each firm has one evening hour and no afternoon hour, these markets of 50 percent and 30 percent, respectively, will be divided equally between the two firms. Thus firm *D* will get 60 percent of the total market if the indicated strategies are used.

Solution to the Marketing Example

Notice that *EEM* dominates the three strategies *AMM*, *MMM*, and *AAA*, while *EAM* dominates *EMM* and *AAM*. This means that firm *D's* last five strategies can be effectively discarded. Once this is done, *AM* dominates *MM*. (If firm *D* plays *AAM*, firm *C* would do better with *MM* than with *AM*; but we have already decided that firm *D* won't play *AAM*.) The game, then, is reduced to one in which each player has five strategies.

One solution to this problem—and, again, I won't say how

I got it—is that firm D should play each of the strategies *EEE, EEA,* and *EAM* one-third of the time, and firm C should play *EE* six-fifteenths of the time, *AA* five-fifteenths of the time, and *AM* four-fifteenths of the time. If firm D uses its recommended strategies, it can be sure of winning on average at least 63 1/3 percent of the audience; and if firm C uses its recommended strategy, firm D won't win any more than this.

Simplified Poker

Figure 3.7

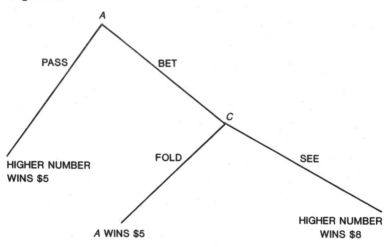

A and C each puts $5 on the table and then toss a coin which has "1" on one side and "2" on the other. Neither player knows the outcome of the other's toss.

A plays first. She may either pass or bet an additional $3. If she passes, the numbers tossed by the two players are compared. The larger number takes the $10 on the table; if both numbers are the same, each gets his or her $5 back.

If *A* bets $3, *C* can either see or fold. If *C* folds, *A* wins the $10 on the table, irrespective of the numbers tossed. If *C* sees, he adds $3 to the $13 already on the table. Again the numbers are compared; the larger number takes the $16, and if the numbers are equal, each gets his or her money back. What are the best strategies, and what should happen?

Each player has four strategies. *A* can always pass (indicated as *PP*), pass with 1 and bet with 2 (*PB*), pass with 2 and bet with 1 (*BP*), or always bet (*BB*). *C* can always fold (*FF*), always see (*SS*), see with 1 and fold with 2 (*SF*), and fold with 1 and see with 2 (*FS*).

To illustrate how the entries in the payoff matrix in figure 3.8 were obtained, suppose *A* plays *BB* and *C* plays *SF*. Half the time, *C* will get 2 and fold (*A* always bets), and *A* will win $5. A fourth of the time, *C* will get 1 and *A* will get 1 and nothing will happen. And a fourth of the time, *C* will get 1 and *A* will get 2; *A* will bet (as always), *C* will see, and *A* will win $8. *A*'s average gain, if these two strategies are adopted is $4.50.

Figure 3.8

	C'S STRATEGIES			
	FF	SS	SF	FS
PP	0	0	0	0
BB	5	0	4½	½
PB	¾	¾	2	0
BP	3¾	−¾	5⁄2	½

A'S STRATEGIES

It can be noticed first that *BB* dominates *BP* and *PP*, while *FS* dominates *SF* and *FF*. The formal model reflects what is probably intuitively clear: whenever *A* gets 2 on the toss, she should bet; and whenever *C* gets 2 on the toss, he should see. The strategy recommended for *A* is to play *BB* three-fifths of the time and *PB* two-fifths of the time. This will assure *A* an average win of 30 cents a play. If *C* plays *FS* three-fifths of the time and *SS* two-fifths of the time, *C* will have a loss of no more than 30 cents, on average, per play.

The Minimax Theorem

Up to now I have considered specific games only. In each of them I recommended certain strategies for the players and indicated what the outcome should be. All these examples were special instances of one of the most important and fundamental theorems of game theory: von Neumann's minimax theorem.

The minimax theorem states that one can assign to every finite, two-person, zero-sum game a value *V*: the average amount that player *I* can expect to win from player *II* if both players act sensibly. Von Neumann felt that this predicted outcome is plausible, for three reasons:

1. There is a strategy for player *I* that will protect this return; against this strategy, nothing that player *II* can do will prevent player *I* from getting an average win of *V*. Therefore, *player* I *will not settle for anything less than* V.
2. There is a strategy for player *II* that will guarantee that he or she will lose no more than an average value of *V*; that is, *player* I *can be prevented from getting away more than* V.
3. By assumption, the game is zero-sum. What player *I* gains, player *II* must lose. Since player *II* wishes to minimize his or her losses, *Player* II *is motivated to limit player* I's *average return to* V.*

The last assumption, which is easily overlooked, is crucial. In non-zero-sum games, where it does not hold, you should

---

*Notice that these three characteristics are almost identical to the ones possessed by games with equilibrium points that were listed in chapter 2.

not conclude that just because player *II* has the power to limit player *I*'s gains, player *I* will necessarily do so. But here, where hurting player *I* is, for *II*, equivalent to furthering his or her own ends, the assumption is compelling.

The concept of mixed strategies and the minimax theorem considerably simplify the study of these games. To appreciate how much, imagine what the theory would be like in their absence. Except for the simplest types of games—games in which there are equilibrium points—there is anarchy. It is impossible to choose a sensible course of action or predict what will happen; the outcome is intimately related to *both* players' behavior and each is at the mercy of the other's caprice. You can only try to define rational play along the lines suggested by Poe, and such attempts are virtually worthless.

With the addition of the minimax theorem, the picture is radically different. In effect, you can treat all two-person, zero-sum games as though they had equilibrium points. The game has a clear value, and either player can enforce this value by selecting the appropriate strategy. The only difference between games with actual equilibrium points and those without them is that in one case you can use a pure strategy and obtain the value of the game with certainty, while in the other case you must use a mixed strategy and you obtain the value of the game on average.

The virtue of the minimax strategy is security. Without it, you must resort to the double- and triple-cross world of Poe's precocious student. With it, you can obtain your full value, and you have the assurance that you couldn't do better—at least, not against good play.

GAME THEORY

Calculating Mixed Strategies

It is considerably harder to calculate mixed-strategy solutions than pure ones. In principle you can find solutions to any zero-sum game, but it may be difficult in practice. The following simple way of finding mixed-strategy solutions *sometimes* works.

To solve a game, first delete all dominated strategies. A strategy may be dominated by either a pure strategy or a mixed one. Domination by a mixed strategy is harder to detect, but a careful inspection of the payoff matrix and a little intuition will usually enable you to eliminate superfluous pure strategies.

In the game shown in figure 3.9, no pure strategy dominates another but your strategy B is dominated by the mixed strategy "play A one-fifth of the time and C, four-fifths of the time" (it is dominated by other mixed strategies as well). If your opponent plays D, you get $1/5(15) + 4/5(5) = 7$, and if your opponent plays E you get $1/5(10) + 4/5(20) = 18$. In both cases, the result exceeds what you would get using B.

Figure 3.9

YOUR OPPONENT

		D	E
	A	15	10
YOU	B	6	15
	C	5	20

Once you eliminate all dominated strategies, you can often find the mixed-strategy solution by following this rule: *Choose a mixed strategy that will give you the same average payoff whatever your opponent does.*

Let's assume that in this game you play A, with probability p, and C, with probability $1-p$, and your opponent plays D, with probability r, and E, with probability $1-r$. If your opponent plays D, your average payoff is $(15)(p) + (5)(1-p) = 5 + 10p$; it is $(10)(p) + (20)(1-p) = 20 - 10p$ if your opponent plays E. If these are the same, $5 + 10p = 20 - 10p$ and p equals $3/4$, the payoff is $12\ 1/2$ on average.

By a similar calculation, you find that your opponent should play each of his or her strategies half the time and the payoff is the same: $12\ 1/2$.

Let's consider another example: A convict has two possible escape routes—the main highway and a forest—and the pursuing sheriff can cover only one of them. If the sheriff and convict take different routes, the escape is certain; if both take the highway route, the convict will certainly be caught. If both go through the forest, the chance of escape is $1 - 1/n$. The matrix entries in figure 3.10 indicate the probability of escaping.

Figure 3.10

		SHERIFF	
		HIGHWAY	FOREST
CONVICT	HIGHWAY	0	1
	FOREST	1	$1-1/n$

Using the same calculations as earlier, we find both sheriff and convict should take the highway with probability $1/(n + 1)$. If either player follows this strategy, the convict will escape with probability $n/(n + 1)$. As n increases and with it the escape probability, the less attractive the forest route becomes to the sheriff; nevertheless that is the route the sheriff must take with increasing probability.

The same method may be applied to larger matrices; in the game depicted on figure 3.11, assume you play strategies A, B, and C, with respective probabilities p, q, and $1 - p - q$ and your opponent plays D, E, and F with respective probabilities, r, s, and $1 - r - s$. Applying the rule, you would conclude that $8p + 6q + 18(1 - p - q) = 6p + 15q + 6(1 - p - q) = 14p + 3q + 6(1 - p - q)$. By playing A, B, and C, with respective probabilities of one-half, one-third, and one-sixth, you would receive an average payoff of 9, whatever your opponent did.

Figure 3.11

YOUR OPPONENT

		D	E	F
	A	8	6	14
YOU	B	6	15	3
	C	18	6	6

Your opponent would obtain a similar pair of equations— $8r + 6s + 14(1 - r - s) = 6r + 15s + 3(1 - r - s) = 18r + 6s + 6(1 - r - s)$—and conclude that he or she should play D, E, and F with respective probabilities four-sixteenths, seven-sixteenths, and five-sixteenths, with the same average payoff of 9.

Solving Two-Person, Zero-Sum Games: A Summary

To solve a two-person, zero-sum game, follow this procedure:

1. Calculate the maximin and minimax—if they are equal, you have found the proper strategies and the value of the game. If they are unequal, go on to step 2.
2. Eliminate all strategies that are dominated.
3. Assign probabilities to each of your strategies so that the outcome of the game will be the same on average whatever your opponent does. Assume your opponent does the same;

if the outcome when you use this mixed strategy is the same as the outcome when your opponent uses his or her mixed strategy and if all probabilities are nonnegative, you have solved the game.

If there is a gap between the outcomes or if some of the probabilities are negative, reexamine the game for dominated strategies; if there are none, then this method failed.

Some Second Thoughts and Applications

Most knowledgeable game theorists, when asked to select the most important single contribution to game theory, would probably choose the minimax theorem. The arguments in favor of minimax strategies are very persuasive; still, they can be overstated. And even a sophisticated writer, in a careless moment, can fall into a trap. In his book *Fights, Games and Debates* (1960), Anatol Rapoport discusses the game shown in figure 3.12. The numbers in the payoff matrix indicate what player *II* pays player *I*. The units are unimportant as long as we assume that both players would like to receive as much as they can.

Figure 3.12

PLAYER *II*

		A	B
PLAYER *I*	a	−1	5
	b	3	−5

Rapoport starts by making the routine calculations. He computes player *I*'s minimax strategy (play strategy *a* four-sevenths of the time and strategy *b* three-sevenths of the time), player II's minimax strategy (play strategy *A* five-sevenths of the time and strategy *B* two-sevenths of the time), and the value of the game (an average win of five-sevenths for player *I*). He then goes on to say: "To try to evade by using a different strategy mixture can only be of disadvantage to the evading side. There is no point at all in trying to delude the enemy on *that* score. Such attempts can only backfire" (p. 160). But Rapoport's conclusion that any deviation from the minimax "can only be of disadvantage" leads to a curious paradox. To see why, observe two basic facts.

Note first that if *either* player selects the minimax strategy, the outcome will be the same: an average gain of five-sevenths for player *I*. There is only one way to obtain another outcome: *both* players must deviate from their minimax strategies.

The second point is that the game is zero-sum. A player cannot win unless he or she wins from the opponent; a player cannot lose unless there is a corresponding gain for the opponent.

Putting these two facts together, we have the contradiction. If a player deviates from the minimax, it cannot possibly be "of disadvantage" unless his or her opponent deviates as well. And, by the same argument, the opponent's deviation must also be of disadvantage. But the game is zero-sum: both players cannot lose simultaneously. Obviously, something is wrong.

Here is the answer. There is a strategy for player *I* that guarantees him or her five-sevenths, and the player can be stopped from getting any more; moreover, player *II* is motivated to stop player *I* from getting more. If player *I* chooses another strategy, he or she is gambling. If player *II* also gambles, there is no telling what will happen. The minimax strategies are attractive in that they offer security; but the appeal of security is a matter of personal taste.

As a consequence of the minimax theorem, the general,

zero-sum, two-person game has a good theoretical foundation. But, like the game of perfect information, it rarely exists in practice. The difficulty is the requirement that the game be zero-sum.

The essential assumption upon which the theory is based is the opposition of the two players' interests. To the extent that the assumption is not valid, the theory is irrelevant and misleading. Often the assumption seems to be satisfied but in reality is not. In a price war, for example, it may be to the advantage of *both* parties that prices be maintained. In a friendly game of poker, *both* players may prefer that neither suffers an excessive loss. And when bargaining, buyer and seller may have divergent interests with respect to the price, but *both* may prefer to reach some agreement. Still, some situations that are best reflected by a zero-sum model—the political arena provides us with a number of them.

Every four years there is a presidential election in the United States. What appears to be one contest is really fifty; each state has its own electoral votes that are won or lost independently from every other state, and the number of electoral votes vary from state to state. The resources with which each party conducts its campaign are money, speeches by notables, media advertising, direct mail, and so forth. These resources are limited for both parties and must be used prudently. The interests of the parties are at opposite poles— there is one president, and if he is to be chosen from one party, he clearly can't be from the other.

Steven Brams and Morton Davis (1974) apply game theory to the problem of allocating resources so as to maximize the number of electoral votes. It is not surprising that candidates tend to spend more of their resources in large states than in small ones, but it seems that they should do so even out of proportion to their size. It is suggested that states should allocate their resources in proportion to the $3/2$ power of the

45

number of electoral votes; a state with four times as many electoral votes as another should get eight times as much resources. An attempt was made to confirm these theoretical results empirically; using campaign appearances as a criterion, it turned out that the "3/2 rule" for allocating resources was a better model for the actual allocation than a proportional model. In a different but similar problem, the same authors consider a different allocation problem: What is the proper way to allocate resources in a sequence of primaries?

The essential difference between this problem and the last one is the time element: a candidate who does well in the early primaries gains political and financial support for the later ones. In this model there is a snowball effect; money spent on the first state in the primary indirectly affects the outcome in all the others. The result is that both in theory and in practice the early primaries are better financed than the later ones. According to the theory, if every state had the same number of electoral votes and there were fifty primaries, a candidate should spend fifty times more on the first primary than it spends on the last one.

The same two authors (1978) analyze another encounter that is best modeled by a zero-sum game: that of two lawyers in a courtroom. It is assumed that there is a population of jurors in which each member has a certain propensity to acquit or convict and that the attorneys can judge this from the potential juror's background. (This is certainly a simplification of the actual situation, but it is generally accepted that there is a correlation between a person's background and his or her actions as a juror.) Both the prosecutor and the defense attorney are given a certain number of peremptory challenges—dismissals of potential jurors without cause—and must decide when to exercise them. The decisions will be based on the attractiveness of the juror being considered and the number of peremptory challenges remaining to both parties. The obvious danger in challenging what appears to be an unfavorable juror is that the next prospect may be worse and the attorney will be helpless.

R. Avenhaus and H. Frick (1976, 1977) describe a zero-sum "game" in a factory. The factory processes material that is either very expensive, such as gold or platinum, or very valuable, such as nuclear fuel. The players are an inspector and a potential diverter. The inspector keeps an inventory of the amount of material that should be there and compares it to the amount of material that is actually there. Although measurements are accurate, they are imperfect; discrepancies are inevitable even when no material is diverted. Certain costs prohibit the inspector from taking inventory too often or setting off too many false alarms, and there are countervailing costs if the inspector fails to discover a diversion or if there is an excessive delay in the discovery. The inspector must observe the discrepancy between book inventory and actual inventory and then balance these costs. The diverter clearly gains from a successful diversion but pays a heavy price if caught. While this is clearly not a strict zero-sum game—the cost to the inspector of excessive inspections is not balanced by a corresponding gain by the potential diverter—it is a close enough approximation to be useful.

In a much less serious vein Robert Bartoszynski and Madan L. Puri (1981) apply a game-theoretical model to a real game: tennis. In tennis, when a player serves, the ball must fall inside certain boundaries; if it fails to do so twice in succession, the player loses the point. If the player succeeds after failing once, there is no penalty for the failure. Most players use a hard first serve, which is difficult to return but somewhat inaccurate, and reserve a more reliable, but less effective, serve if the first serve fails. Bartoszynski and Puri investigated under what conditions a player should use the hard serve both times, neither time, or just the first time.

The authors also address the following problem: suppose you have been preparing a special serve to use against your usual tennis partner. The service is to be a surprise that will only be effective once—a slice that bounces deceptively, perhaps. At what point in the game should you use it? Bartoszynski and Puri's somewhat surprising conclusion is that it makes no difference when you use it—the effect will be exactly the same. The important thing is that you use it at some time. If you reserve its use for a situation that may not arise—only when you are leading 40 to 15, for example—you lose part of your advantage.

Another, surprising application to game theory was made some time ago by mathematician Edwin O. Thorpe and W. E. Waldman (1973). People have been gambling for a very long time, and it is very likely that, for as long a time, there have been players with systems. Some systems are based on lucky numbers, omens, using past history to predict the future, changing the amounts bet, and so forth. But since the house fixed the odds so that each individual bet was unfavorable, no sorcery could alter the inevitable loss to the longtime player. So systems players patronized the gambling houses and the gambling houses welcomed and thrived on them. Until the appearance of Edwin Thorpe. Thorpe realized that making a profit on a sequence of bets when you lose on average on each one is like losing money on each sale and trying to make it up in volume. He also realized that in a game like blackjack, the situation isn't really repeated unless the cards are dealt thoroughly after each deal. Thorpe, with the aid of a computer, devised a number of techniques for determining whether the cards played in the past improved a player's chances. Thorpe's methods were described in a bestselling book, *Beat the Dealer*, (1966) and gambling casinos tacitly acknowledged his success: they changed some of the rules, combined decks to make counting harder, and attempted to bar card counters from the tables. In a more recent paper written jointly with William E. Walden, Thorpe (1973) analyzes such games as baccarat, poker, and trente-et-quarante in which cards are exposed without being redealt and in which this information can be put to a player's advantage.

Another application of game theory to real games was made by Nesmith C. Ankeny (1981). Poker has long been considered a fertile area for applying game theory, and a number of papers have been written on the subject, including a substantial part of *Theory of Games and Economic Behavior* by von Neumann and Morgenstern (1953). But generally, only simplified forms of poker were analyzed. Ankeny, a mathematics professor at M.I.T., addresses the game of poker as it is actually played in his book *Poker Strategy: Winning with Game Theory.* He suggests a "proper balance of deception and strength" to maximize winnings. Certain aspects of bridge, such as signaling, have also been analyzed, and mixed strategies have been applied to bridge play; but no overall model, encompassing both bidding and play, has been devised.

Military tactics, an area in which the parties are involved in almost pure conflict, is another source of two-person, zero-sum games. Evasion and pursuit games and fighter-bomber duels are two applications. In addition, there is a general class of games called Colonel Blotto games, in which the players are required to allocate certain resources at their disposal. We have seen one example of this in a military context—Generals X and Y allocating their divisions—and another in a marketing context. The same problem arises when allocating salespeople to various territories or police officers to high-crime areas. An interesting application of Colonel Blotto games arose in the process of creating a model for inspection in a possible disarmament agreement. One country has factories that produce certain types of weapons—some more potent than others, some easier to hide than others, and so on. The inspecting country has to decide how to allocate its inspectors to verify that the inspected country is not exceeding its quota of weapons.

An interesting application of minimax strategies was observed by Sidney Moglower (1962). In the process of selecting crops, the farmer can be regarded as one player and the "hypothetical combination of *all* the forces that determine market prices for agricultural products" as the other player. Moglewer points out that though it is difficult to justify the farmer's implicit assumption that the universe is concerned with him or her as an individual, the farmer acts as though this were the case.

Some Experimental Studies

Although I speak of game *theory*, it is important to look at games in practice. In a speech to future jurists at the Albany Law School, Justice Benjamin Cardozo remarked, "You will study the life of mankind, for this is the life you must order, and to order with wisdom, know." The comment applies to game theory as well. Game theory has its roots in human behavior; if the theory is not related to human behavior in some way, it will be sterile and meaningless except as pure mathematics. Aside from other considerations, experiments in game theory are interesting in themselves, perhaps because people enjoy reading about what other people do. This is sufficient reason for discussing them.

Richard Brayer (1964) had subjects repeatedly play the game shown in figure 3.13. The entries in the payoff matrix indicate what the subject received from the experimenter.

The players either were told they were playing against an experienced player or they were told that their opponent would play randomly, but in fact they invariably played against the experimenter.

Figure 3.13

		EXPERIMENTER		
		A	B	C
	a	11	−7	8
SUBJECT	b	1	1	2
	c	−10	−7	21

It takes only a moment for subjects to deduce that they should play strategy *b* if they feel their opponent is intelligent. From the point of view of the experimenter, strategy *C* is dominated by *B*, so subjects can assume that the experimenter will never play *C*. Having decided this, subjects should never play *c*, since they always do better with strategy *b*. Eliminating *c* and *C*, the experimenter should play *B*, and subjects should play *b*. (*B*, *b*) are equilibrium strategies, and the payoff 1 is an equilibrium point.

What actually happened? For one thing, the subjects ignored what they were told about their opponent and responded only to how the opponent played. Whether they did not see the relevance of what they were told about their opponent, or did not know how to apply it, or whether they were reacting with the natural skepticism subjects have toward anything they are told by an experimenter is not clear. The subjects did play *b* if the experimenter played *B*, but not otherwise. Postexperimental interviews confirmed what the play indicated: the players couldn't anticipate the experimenter's choice of strategy *B*. In fact, *more than half the subjects felt that the experimenter was stupid for playing* B *and settling for a loss of 1*. When the experimenter picked a strategy at random, the subjects generally responded by playing strategy *a:* the one that gave them the highest average return.

The same pattern was observed by other experimenters. Several—Theodore Caplow (1956, 1959), Merrill M. Flood (1952), Oliver L. Laccy and James L. Pate (1960), Bernhardt Lieberman (1960), Robert E. Morin (1960), and Anatol Rapoport and Carol Orwant (1962)—concluded that most subjects are simply unable to put themselves in the shoes of their opponents. Subjects tend to prefer strategies that yield an apparently high average return (against an opponent playing at random) to equilibrium strategies. When players are punished for deviating from the equilibrium strategy, they change their behavior, but not otherwise.

In games without equilibrium points, the players have even less insight. In an interesting review of experimental games, Anatol Rapoport and Carol Orwant (1962) suggest that equilibrium points, when they exist, will be found eventually, if not immediately, by the players. The speed with which they are discerned varies with the experience and sophistication of the players and the complexity of the game. When there are no equilibrium points, when what is called for is a mixed strategy, the situation is much worse. Not only are all but the most sophisticated players incapable of making the necessary calculations, but almost all players do not even see the need for them.

Is the failure to play an equilibrium strategy, where one exists, necessarily proof of a player's ignorance? Perhaps not on the face of it. It might be argued that the case for the equilibrium strategy is based to some extent on the assumption that one's opponent will play reasonably well; if this is not true, a player often does better by not playing the equilibrium strategy. When the experimenter played the equilibrium strategy, the subjects did too. When the experimenter played randomly and "irrationally," however, the subjects persisted in playing nonequilibrium strategies.

All this sounds plausible but, as mentioned earlier, it isn't what happened. By their own admission after the experiments, the players indicated that they had no insight into what was happening. They "learned" how to react effectively to the specific behavior of the experimenter (all the while believing the equilibrium strategy the experimenter chose was foolish), but the subjects' ideas of what the game was about were as muddy at the end as they were at the start.

Apropos of "learning" how to play games, there is a theorem that is interesting enough to warrant a digression. Suppose that two people are playing a game repeatedly but do not have the skill to compute the minimax strategies. They both play randomly the first time, and subsequently they learn from their experience, in the following sense. They each assume that their opponent is playing a mixed strategy with probabilities proportional to the actual frequencies with which strategies have been chosen in the past. On the basis of this assumption, they play the pure strategy that will maximize their average return. Julia Robinson (1951) proved that, under these conditions, both strategies will approach the minimax strategies.

Of what practical significance are these experiments? Does it make any difference to a player to know that players often act irrationally? It didn't matter in games of perfect information such as chess. Does it matter now?

It is always possible to get the value of the game by playing minimax. That is so whether one's opponent is rational or not. The reason players are satisfied to get the value of the game is that they know a clever opponent can stop them from getting any more. But if players have reason to believe that their opponent will play badly, why not try to do better?

By using the minimax strategy, players will avoid doing

anything foolish, such as playing a dominated strategy. In our simplified poker game, players would always bet if they were dealt a 2; in the marketing game, they would never select three morning hours to advertise; and, in the military game, they would never send five divisions to one location. It should be emphasized, however, that the minimax strategy is essentially defensive, and when you use it, you often eliminate any chance of doing better. In Poe's odd-even game and in the game discussed by Rapoport, if a player adopts the minimax strategy, he or she obtains the value of the game; no less, but *no more.*

Suppose that, in the odd-even game, your opponent tends to choose even more often than he or she should. How can you exploit this if the game is played only once and you don't know how your opponent plays? Even if you know the game is so complex that he or she is not likely to analyze it correctly, how can you anticipate in what direction your opponent will err? In general, you may suspect that your opponent will pick the strategy that seems to yield the largest average payoff; but if you act on your suspicion, you become vulnerable to an opponent who is one step ahead of you and who is out to get rich by exploiting the exploiter. And if the game is played repeatedly and you manage to do better than you should, your opponent will eventually learn to protect himself or herself, so your advantage is unstable.

The weakest part of the theory is undoubtedly the assumption that a player should always act so as to maximize the average payoff. The justification for the assumption is that, in the long run, not only the average return but the actual return will be maximized. But if a game is played only once, are long-run considerations relevant? John Maynard Keynes (1972) made the point in *Monetary Reform:* "Now 'in the long run' this is probably true. . . . But this *long run* is a misleading guide to current affairs. *In the long run* we are all dead." Often a strategy that maximizes the average return is not desirable, much less compelling. Is it irrational to opt for a sure million dollars rather than take an even chance of getting $10 million? This is not just an incidental objection; it goes to the heart of the matter. Let's take a closer look at this problem. In the game in figure 3.14, the payoffs are in dollars.

Figure 3.14

PLAYER *II*

		A	D
	a	$1 million	$1 million
PLAYER *I*	b	$10 million	$0
	c	$0	$10 million

The minimax strategy for player *I* is to play *b* and *c*, with a probability of one-half. Using this strategy, player *I* will get $10 million half the time and will get nothing half the time. But player *I* may prefer to have a sure million; in fact, rather than gamble, both players may prefer that player *I* win a sure million. This is one reason why cases are often settled out of court.

Things are not always what they seem. We started with what appeared to be a zero-sum game (it was a zero-sum game in dollars), but in fact the players had certain common interests. Yet the critical assumption in zero-sum games is that the *players have diametrically opposed interests*. And this must be so not only for the entries in the payoff matrix, which indicate the outcome when each player uses a pure strategy, but also for the various probability distributions that may be established when the players use mixed strategies.

This objection is almost fatal. It can be overcome only by introducing an entirely new concept, the concept of utility. (The word is old, but the idea is new.) The utility concept puts the theory on a firm foundation once again and in fact is one of von Neumann and Morgenstern's most significant contributions to game theory. It will be discussed next.

Solutions to Problems

1. If your coin turns up 1, you should always bid. If your coin turns up 0, you should bid (bluff) one-eleventh of the time and pass the remaining ten-elevenths of the time.

    If your opponent's coin turns up 1, he or she should always see. If your opponent's coin turns up 0, he or she should see one-eleventh of the time and fold the remaining ten-elevenths of the time.

    You should win $25/11 on average each time you play.

2. The police officers should never split up; they should go to the airport half the time and to the seaport the other half. The smugglers should split up four-fourteenths of the time; both should go to the airport five-fourteenths of the time and to the seaport five-fourteenths of the time. Fifty pounds of contraband will get through on average.

3. Ninety percent of the time the watchman should check the safe with $90,000, (which is what one might suspect), but 90 percent of the time the burglar should try to rob the safe with $10,000 (which is not so clear). On average the burglar should get $9,000.

4. You should bid the slam with probability of .1 and your opponent should bid the slam with a probability of .9. In fact, the *less* likely the slam is to succeed the more likely your opponents will bid it. In this game your chance of winning the tournament is .91. (which is slightly better than the .9 you would get if you always bid game and your opponent always bid slam).

4

Utility Theory

Introductory Problems

Game theory is a tool for making decisions; but before you decide how to get what you want, you must first decide what you want. Sorting out what you want—the function of utility theory—is not always as easy as it sounds. Before reading the chapter, think about what you would do in each of the following situations.

1. When you arrive at a play, you find you have lost your pair of $40 tickets. Would you spend another $40 to buy a new pair of tickets or look elsewhere for your evening's entertainment?
2. Which would you prefer: a sure million dollars or an even chance at $3 million?
3. A rich aunt dies and leaves you $200; soon after a sporting uncle offers you the choice of another $50 or a 25 percent chance of winning $200. Which do you accept?
4. Suppose that 1 percent of the people of your age and health die in a given year. How much would you pay for $100,000 worth of life insurance?
5. Would you prefer a sure $10 or an even chance of $30?

57

6. You arrive at a theater expecting to pay $40 for tickets and find you have lost $40 of your money. You still have more than $40 left—would you buy the tickets anyway?

7. A rich aunt (not the one in question 4) dies and leaves you $400. You are caught speeding and a whimsical judge offers you the following choice: You can pay a $150 fine, or you can pay $200 with a 75 percent probability and pay nothing with a 25 percent probability. What do you do?

The last chapter was primarily concerned with the minimax theorem—why it is needed, what it means, and why it is important. For the sake of clarity, I discussed the central ideas and avoided mentioning the difficulties. At the end of the chapter, however, it was suggested that all was not as it should be. Part of the foundation, and an absolutely essential part, is missing, and the theory is in fact firmly rooted in midair. One of von Neumann and Morgenstern's important contributions is the concept of utility, which makes the old solutions and strategies plausible once again. The utility function was fashioned precisely for this purpose; and in this chapter I will discuss utility functions and their role in game theory.

The heart of the difficulty is the ambiguity of the term "zero-sum." Taking the simplest view, a game is zero-sum if it satisfies a certain conservation law: a game is zero-sum if, during the course of the game, wealth is neither created nor destroyed. In this sense, ordinary parlor games are zero-sum.

But this definition won't do, for it is not the payoffs in money that are important. Throughout the discussion of the two-person, zero-sum game, it was assumed that each player was doing his or her best to hurt an opponent; if this assumption fails, the rest of the theory fails with it. What is needed to make the earlier argument plausible is the assurance that the players will indeed compete. A parent playing a card game with a child for pennies is playing a zero-sum game in my original sense, yet the arguments of the last chapter do not apply. Of course, if we *assume,* as we did earlier, that a player's goal is to maximize the expected payoff in money, there is no problem, but this begs the question; it is the validity of this assumption that is in doubt.

As a matter of fact, there are many situations in which people *don't* act so as to maximize their expected winnings. This has nothing to do with the formal theory; it is an observation about life. The game used as an illustration at the end of the last chapter is only one example in which the assumption is incorrect; there are many others.

Generally, a person's wealth affects his or her attitude toward risk. A multimillion-dollar corporation will take a $50,000 gamble when a person with exactly that much capital will not. And this is so even when both perceive the situation in exactly the same way. Eli Schwartz and James Greenleaf (1978) suggest that this is one reason that the rich grow richer while the poor grow poorer. They constructed a model of a society in which everyone is equally wealthy initially and in which there was a sequence of bets that each could make. Some bets were riskier than others but their average return was higher as well. As chance made some members wealthier and others poorer, this difference was accentuated in a systematic way since the more affluent could afford the more speculative, higher-return bets. Schwartz and Greenleaf found in some computer runs that after five, twenty, and fifty trials the top tenth controlled 18 percent, 25 percent, and 50 percent of the total wealth, respectively.

Many games having a negative average payoff nevertheless attract a large number of players. For instance, people buy lottery tickets, play the "numbers game," bet at the race track, and gamble in Atlantic City, Las Vegas, and Reno. This isn't to say that people bet aimlessly without pattern or purpose; certain bets are generally preferred to others. In studies of parimutuel betting at the race track, for instance, Griffith (1949) observed that bettors consistently gave more favorable odds on long-shots than experience warranted, while the favorites, at more attractive odds, were neglected. Just what it is that makes one bet more attractive than another is not always clear; what is clear is that the players are trying to do something other than maximize their average winnings.

It isn't just the gamblers who are willing to play games with unfavorable odds; such games are also played by conservatives who wish to avoid large swings. Commodity-futures markets are created because at planting farmers want to be sure prices don't drop too low by harvest. Also, almost every adult carries personal insurance of one sort or another, yet the "game" of insurance has a negative value, for the premiums cover not only the benefits of all policyholders but insurance company's overhead and commissions as well. Incidentally, insurance is just a lottery turned around. In both cases, the player puts in a small amount; lottery players have a small chance to win a fortune, and insurance policyholders avoid the small chance of having a catastrophe befall them.

The prevalence of insurance policies reflects willingness to pay a price for security; this aversion to risk has also been observed in the laboratory.

Harry Markowitz (1955) asked a group of middle-class people whether they would prefer to have a smaller amount of money with certainty or an even chance of getting ten times that much. The answers he received depended on the amount of money involved. When only a dollar was offered, all of them gambled for ten, but most settled for a thousand dollars rather than try for ten thousand, and all opted for a sure million dollars.

It should be emphasized that the failure to maximize average winnings is not simply an aberration caused by a lack of insight. In 1959 Alvin Scodel, J. Sayer Minas, and Philburn Ratoosh, asked the subjects of an experiment to select one of several bets. There was no correlation whatever between the bets selected and the subject's intelligence. Graduate students in mathematics were included among the subjects, and they certainly could do the necessary calculations if they wished.

So if the theory is to be realistic—and if it is not realistic, it is nothing—we cannot assume that people are concerned only with their average winnings. In fact, we cannot make any general assumption about people's wants, because different people want different things. What is needed is a mechanism that relates the goals of a player, whatever they are, to the behavior that will enable him or her to reach these goals. In short, a theory of utility.

Before you can make sensible decisions in a game, both the goals of the players and the formal structure of the game must be taken into account. In this process of decision making, there is a natural division of labor. The game theorist, to paraphrase Lewis Carroll, must pick the proper road after being told the player's destination; and players don't have to know anything about game theory, but they have to know what they like—a slightly altered version of the popular bromide, "I don't know anything about art but I know what I like."

The problem is to find a way for players to convey their attitudes in a form that is useful to the decision maker. Statements such as "I detest getting caught in the rain" or "I love picnics" do not help someone decide whether to call off a picnic when the weather prediction is an even chance of rain. There is no hope of completely describing subjective feelings quantitatively, of course, but, using utility theory, it is possible to convey enough of these feelings (under certain conditions) to satisfy my present purpose.

61

GAME THEORY

Utility Functions: What They Are, How They Work

A utility function is simply a "quantification" of a person's preferences with respect to certain objects. Suppose I am concerned with three pieces of fruit: an orange, an apple, and a pear. The utility function first associates with each piece of fruit a number that reflects its attractiveness. If the pear was desired most and the apple least, the utility of the pear would be greatest and the apple's utility would be least.

The utility function not only assigns numbers to fruit; it assigns numbers to lotteries that have fruit as their prizes. A lottery in which there is a 50 percent chance of winning an apple and a 50 percent chance of winning a pear might be assigned a utility of 6. If the utilities of an apple, an orange, and a pear were 4, 6, and 8 respectively, the utilities would reflect the fact that the person was indifferent (had no preference) between a lottery ticket and an orange, that he or she preferred a pear to any other piece of fruit or to a lottery ticket, and that he or she preferred a lottery ticket or any other piece of fruit to an apple.

Also, utility functions assign numbers to all lotteries that have as prizes tickets to other lotteries; and each of the new lotteries may have as its prizes tickets to still other lotteries, so long as the ultimate prizes are pieces of fruit.

This is still not enough, however. Von Neumann and Morgenstern demand one more thing of *their* utility functions that make them ideally suited for their theory. The utility functions must be so arranged that the utility of any lottery is always equal to the weighted average utility of its prizes. If in a lottery there is a 50 percent chance of winning an apple (which has a utility of 4) and a 25 percent chance of winning either an orange or a pear (with utilities of 6 and 8, respectively), the utility of the lottery would necessarily be $5\frac{1}{2} = (.5)(4) + (.25)(6) + (.25)(8)$.

The Existence and Uniqueness of Utility Functions

It is easy enough to list conditions that you would like the utility functions to satisfy; it is quite another thing to find a utility function that does. Given a person of arbitrary tastes, is it always possible to find a utility function that reflects them? Can these tastes be reflected by two different utility functions?

Taking the second question first, the answer is yes. Once a utility function is established, another can be obtained from it by simply doubling the utility of everything; and still another, by adding one to the utility of everything. In fact, if we take any utility function and multiply the utility of everything by any positive number (or add the same number to every utility), a new utility function is obtained that works just as well.

It is also possible that there is no utility function that will do; this would be the case if the person's tastes lacked "internal consistency." Suppose a person, when given a choice between an apple and an orange, prefers the apple. That would mean that the utility of an apple would have to be greater than the utility of an orange. If, between an orange and a pear, the person prefers the orange, the utility of an orange would have to be greater than that of a pear. Since the utilities are ordinary numbers, it follows that the utility of an apple is necessarily greater than the utility of a pear. If, when given the choice between an apple and a pear, the person chooses the pear, establishing a utility function would be hopeless. There is no way one can assign numbers to the pieces of fruit that will simultaneously reflect all three of these preferences. There is a technical name for what I have just described: the player's preferences are said to be *intransitive*.

If the preferences of a player are sufficiently consistent— that is, if they satisfy certain requirements—the preferences can be expressed concisely in the form of a utility function.

GAME THEORY

Six Conditions That Guarantee the Existence of a Utility Function

If a player's preferences are to be expressed by a utility function, these preferences must be consistent; that is, they must satisfy certain conditions. These conditions may be expressed in several more or less equivalent ways; I have used the formulation suggested by Duncan R. Luce and Howard Raiffa (1957). For convenience, the word "object" will represent either a piece of fruit or a lottery.

1. *Everything is comparable.* Given any two objects, the player must prefer one to the other or be indifferent to both; no two objects are incomparable.
2. *Preference and indifference are transitive.* Suppose A, B, and C are three different objects. If the player prefers A to B and B to C, A will be preferred to C. If the player is indifferent between A and B, and B and C, he or she will be indifferent between A and C.
3. *A player is indifferent when equivalent prizes are substituted in a lottery.* Suppose in a lottery one prize is substituted for another but the lottery is left otherwise unchanged. If the player is indifferent between the old and new prizes, he or she will be indifferent between the lotteries. If the player prefers one prize to the other, he or she will prefer the lottery that offers the preferred prize.
4. *A player will always gamble if the odds are good enough.* Suppose that, of three objects, A is preferred to B and B is preferred to C. Consider the lottery in which there is a probability p of getting A and a probability of $1 - p$ of getting C. Notice that if p is zero the lottery is equivalent to C, and if p is 1, the lottery is equivalent to A. In the first case the lottery is preferable to B, while in the second case B is preferable to the lottery. According to this condition, that there is a value p between zero and 1 that will make the player indifferent between B and the lottery.
5. *The more likely the preferred prize, the better the lottery.* In lotteries *I* and *II* there are two possible prizes: objects A

and *B*. In lottery *I*, the chance of getting *A* is *p*; in lottery *II*, the chance of getting *A* is *q*. *A* is preferred to *B*. This condition requires that if *p* is bigger than *q*, lottery *I* be preferred to lottery *II*; and, conversely, if lottery *I* is preferred to lottery *II*, *p* is greater than *q*.

6. *Players are indifferent to gambling.* A player's attitude toward a compound lottery—a lottery in which the prizes may be tickets to other lotteries—is dependent only on the ultimate prizes and the chance of getting them as determined by the laws of probability; the actual gambling mechanism is irrelevant.

Henceforth it will be supposed that the preferences of each player are expressed by a utility function and that the payoffs are given in "utiles"—the units in which utility functions are expressed. When I speak of a zero-sum game now, I mean a game in which the sum of all the payoffs (in utiles) is always zero. When two players are involved in a zero-sum game in this new sense, their interests must necessarily be opposed.

The advantage of this new definition of zero-sum is that, if applied to the new zero-sum games, it justifies the work earlier on two-person, zero-sum games. But, admittedly, there is a corresponding disadvantage in the new approach. This might have been anticipated as a consequence of our quasi-conservation law: games that are easy to analyze don't come up very often. The difficulty is that this new type of zero-sum game is rare and hard to recognize. Before I changed the definition, you only had to know the rules of a game such as poker to see immediately that it was zero-sum. Now the problem of recognition is much more subtle and involves subjective factors such as the players' attitude toward risk. This makes the theory much more troublesome to apply.

GAME THEORY

Some Potential Pitfalls

Utility theory is easily misunderstood. The reason for this is partly historical: "utilities" have been in existence for some time, but the term has not always been used in a consistent or clear way. In an article called "Deterrence and Power," Glenn Snyder (1960) gave a number of good examples of how one can go wrong, as follows:

> The figures which follow [i.e., the payoff matrix] are based on the assumptions that both sides are able to translate all relevant values into a "common denominator" "utility," that they can and do estimate probabilities for each other's moves, and that they act according to the principle of "mathematical expectations." The latter states that the "expected value" of any decision or act is the sum of the expected values for all possible outcomes, the expected value for each outcome being determined by multiplying its value to the decision-making unit times the probability that it will occur. To act "rationally" according to this criterion means simply to choose from among the available courses of action the one which promises to maximize expected value (or minimize expected cost) over the long run. There are reasons—apart from the practical difficulty of assigning numerical values to the elements involved—why the mathematical expectation criterion is not entirely suitable as a guide to rationality in deterrence and national security policy. However, it is useful as a first approximation, the necessary qualifications—having to do with the problem of uncertainty and the disutility of large losses—will be made presently. (p. 168)

Snyder starts out all right; you do have to "translate all relevant values into a 'common denominator' 'utility.'" That this is possible must be assumed, but that is *all* you have to assume. Snyder is also willing to assume that "they act according to the principle of 'mathematical expectations.'" But why should a person, or a country, want to maximize its utility, any more than it would want to maximize its average winnings in money?

Both the last question and Snyder's assumption put the cart before the horse. The person's wants come first; the utility function, if it exists, comes second. The person is *not* trying to maximize his or her utility—average or otherwise; very likely, the person doesn't even know such a thing exists. Players may act *as though* they were maximizing their utility function, but this is not because they mean to but because of the judicious way the utility function was established. What happens, at least in theory, is that the preferences of the players are observed and then a utility function is established that the players *seem* to be maximizing.

Snyder's statement "To act 'rationally' according to this criterion means simply to choose from among the available courses of action the one which promises to maximize expected value (or minimize expected cost) over the long run" can be criticized on two counts. For one thing, as with the earlier statement, the problem is put the wrong way around. You don't maximize your utility because you're rational; your preferences are observed *first* and *then* the utility function is established. Moreover, "over the long run" has nothing to do with it. Utilities are assigned to one-shot gambles. If a person takes the view that life is a series of gambles in which good and bad luck cancel each other, she may decide to maximize her average return in dollars; and this is all very well. But if she prefers less risky alternatives, that's all right too, even if it lowers her expected return. Utility theory will accommodate either attitude. In the deterrence "game," gambles are taken without any assurance that they will recur; in fact, there is good reason to believe they will not.

And, finally, Snyder's reservations about the "disutility of large losses" entirely misses the point. Either a utility function exists or it doesn't. If the utility function associates with large losses numbers that are inaccurate, then it isn't a utility function at all. A utility function, if it does anything, must reflect a person's preferences accurately; it serves no other purpose.

Constructing a Utility Function

Granted that there is a need for a utility function, there still remains the problem of actually determining it. Roughly speaking, this is how it can be done. People are asked to make many simple choices between two things; it may be pieces of fruit, or lottery tickets. People might be asked, for example, whether they would prefer a ticket to a lottery in which there is a three-fourths chance of winning a pear and a one-fourth chance of winning an apple, or the certainty of getting an orange. On each question they must indicate a preference for one of the alternatives or state that they are indifferent between both. It is possible, on the basis of these simple choices, to establish a single utility function that assigns a number to every piece of fruit and every possible lottery—reflecting all the preferences of an individual simultaneously, providing players' preferences are consistent. In effect, the fruit and lotteries are put on a single dimension in which every piece of fruit and every lottery ticket can be compared simultaneously.

It should be understood that in the process of establishing a utility function, nothing new was added; the grand final ordering is implicit in the simple choices made earlier. But the practical advantage of having a concise utility function rather than a great many individual preferences is enormous.

Are People's Preference Patterns Really Consistent?

So far I have assumed that people for whom utility functions are to be established have consistent preferences; that is, preferences that satisfy the six consistency conditions listed earlier. On the face of it, these conditions seem intuitively appealing, and one would think that most people would accept them with few if any objections. It has even been suggested that these conditions—or, at any rate, some of them—should be used as a *definition* of rationality in decision making. But it turns out that people often have, or seem to have, inconsistent attitudes that preclude the construction of a utility function. Let's look at some of the things that can go wrong.

A fair amount of effort has been expended to study betting behavior experimentally and, in particular, to see whether it is consistent. In a number of experiments, Ward Edwards (1953, 1954 *a, b, c, d*) compared the bets that subjects chose with certain other variables, such as the average amount the subject could win, the state of the subject's fortunes when the choice was made, the actual numerical probabilities involved, and so forth. He found, among other things, that certain subjects tended to make bets in which the probabilities involved were one-half, one-half rather than one-quarter, three-quarters, even though the average return was the same in both cases. It is easy to show that a person cannot consistently prefer the one type of bet to the other and still satisfy the six conditions necessary for consistency.

Another source of difficulty is the variation in people's preferences over a period of time. This might not appear to present too serious a problem, since variations would occur gradually; but that is not the case. The interaction between what happens in the game and the attitudes of the players cannot be neglected at any time. A worker who has been out on strike for a long time looks at an offer with different eyes than a worker just starting to negotiate; and, during the process of negotiation, flexible attitudes may harden and what were once acceptable possibilities may become unacceptable.

Anyone who has watched people gambling will no doubt have observed the same thing. In a poker game, for example, as time passes, the bets grow larger: apparently the amount that a player considers an acceptable risk increases as the game proceeds. The same phenomenon has been observed experimentally by comparing the size of bets made in the early and the late races at the track. A subjective but graphic picture of what happens is given by Dostoevsky (1915) in his novel *The Gambler.*

> As for me, I lost every farthing quickly. I staked straight off twenty friedrichs d'or on even and won, staked again and again won, and went on like that two or three times. I imagine I must have had about four hundred friedrichs d'or in my hands in about five minutes. At that point I ought to have gone away, but a strange sensation rose up in me, a sort of defiance of faith, a desire to challenge, to put out my tongue at it. I laid down the largest stake allowed—four thousand gulden—and lost it. Then getting hot, I pulled out all I had left, staked it on the same number, and lost again, after which I walked away from the table as though I were stunned, I could not even tell what happened to me.... (p. 168)

There are many other problems as well. Experiments have shown that decisions often depend on seemingly irrelevant variables. People make one type of bet when playing for money and another when playing for wooden chips that are worth money. They bet one way when others are present and another way when they're alone. Their history—the success they've had in the game so far—influences their attitude toward risk. People select inconsistently; they pick one of two bets one time, and the other the next. And their preferences are intransitive: they prefer A when offered a choice between A and B, and B when offered a choice between B and C. And when offered a choice between A and C, they prefer C.

Note how the context of a problem, as described by Leonard J. Savage (1954), can influence decision making: "A man buying a car for $2,134.56 is tempted to order it with a radio installed, which will bring the total price to $2,228.41, feeling that the difference is trifling. But when he reflects that, if he already had the car, he would certainly not spend $93.85 for a radio for it, he realizes that he had made an error" (p. 103).

Recently serious doubts were raised about the foundations of utility theory. In an experiment conducted by Daniel Kahneman and Amos Tversky (1982), it was found that most subjects made different decisions in two identical or equivalent situations when these situations were described in different ways. The general rule was this: people who feel they have won something generally try to conserve their winnings by avoiding risks. In an identical situation, the same people who perceive that they have just lost something will take risks they considered unacceptable before, to make themselves whole. (See "Solutions to Problems," p. 73, for more details on this experiment.)

In the same experiment, subjects were told a jacket cost $125 and a calculator $15. But they were offered a chance to get the calculator for $10 if they drove twenty minutes to another store. Most accepted. But when the prices were switched and $5 was saved on the $125 item instead of the $15 item, most subjects declined to travel; presumably a decrease of 4 percent was considered less significant than a decrease of 33⅓ percent. But in fact this is another variation of Savage's paradox, because in both cases you save $5 by traveling 20 minutes.

Despite all the difficulties—despite the seeming irrationality of people's behavior, despite the inconsistencies—utility functions have been established successfully. In one case experimenters observed the bets that subjects chose when they were offered simple alternatives. On the basis of these observations, they were able to predict what the subjects would do when presented with much more complicated decisions. How was this possible, in view of all the objections mentioned?

Not all the objections can be explained away, of course, but the problem is not as serious as it may seem. Take the case of intransitive preferences, for instance. Some experimenters feel that true intransitivities rarely arise. What happens, they say, is that people are forced to choose between alternatives to which they are indifferent. The answer, then, is dependent on momentary whim. If people have difficulty deciding whether to buy a car for $7,000, it should not be surprising if one moment they refuse to buy at a dollar less and the next moment they agree to buy at a dollar more. In several experiments it was observed that subjects who make intransitive choices don't make them consistently. In one experiment, for instance, subjects were asked to list certain crimes in the order of their seriousness and to determine the severity with which guilty parties should be punished. While occasional intransitivities turned up, when the task was repeated they almost invariably disappeared.

It is likely that much "irrational behavior" can be avoided by identifying and controlling the significant variables. If, for example, people bet more aggressively when they are with others than when they are alone (and they do), this must be taken into account. Also, the subjects should have some experience in decision making, so they can take into consideration the consequences of their actions. In Savage's example, this is precisely what was needed.

Finally, people must be properly motivated. In some experiments subjects stated that they had varied from what they believed to be the most prudent course, just to keep things exciting. This is a serious problem, since as a rule experimenters have limited resources and cannot pay enough to motivate the players to the degree they would like. People betting in a "hot, humid, exciting, and highly disordered carnival setting," as one experimenter put it, behaved quite differently from bettors in an artificial, insulated, experimental environment.

Solutions to Problems

1, 3, 6, and 7. The point of these questions is that people do not always make decisions based on their situations but rather on how their situations arose or were described. You are more likely to buy tickets in the situation of question 1 than in question 6 if you are like the subjects of an experiment by Kahneman and Tversky (1982). Also, most subjects declined the gamble offered in question 3 but accepted the gamble in question 7 although the two cases were identical: a sure $250 versus a 25 percent chance at $400 and a 75 percent chance of $200. This apparent inconsistency raises doubts about the foundations of game theory.

When people are faced with a loss, they seem to seek out risks—they want to gamble to remain whole. The loser in a poker game who wants to play double or nothing and the holder of a losing stock waiting to break even are two examples of this kind of behavior. People who are winners often want to avoid risk—witness the winning poker player who wants to go home. This seems to account for the difference in behavior because the objective situations are identical.

2, 4, and 5. Game theory does not tell you whether you should gamble or not—it tells you how to achieve what you want. There is nothing irrational about preferring a sure million dollars in question 2 and gambling for $30 in question 5—in fact, most people would do the same. The "fair" price of insurance is $1,000 in question 4 (this would mean you would break even if you took this bet many times), but there is no reason you can't pay more if you want insurance and no reason you can't decline a lower premium if you don't want insurance.

5

The Two-Person, Non-Zero-Sum Game

Introductory Problems

The games I have been discussing until now have been zero-sum—your gain was your opponent's loss. The non-zero-sum games that we encounter are at once more commonplace, more interesting, and more difficult to analyze. There are theories that favor one strategy over another but the arguments are less persuasive; the ultimate outcome can be conjectured about but the predictions are less compelling. The problems in this and the final two chapters have more to do with ideas than numbers. For this reason prethinking about the introductory ideas—before the text shapes your ideas—will be particularly rewarding.

GAME THEORY

1. You may assume that the
   matrix entries in figure 5.1
   are in dollars. With each pair
   of strategies there are *two*
   payoffs: one for you (shown
   first), and one for your partner. If you choose B and your

Figure 5.1

partner, A, you would get $5 and your partner would get
   nothing. Communication between players is not allowed.
   a. What strategy would you choose if this game were
      played only once?
   b. Would you play any differently if the game was to be
      repeated 200 times?
   c. Some time ago a number of people were asked to write a
      computer program describing exactly how to play on
      each of 200 repetitions of the game; your play on the
      fiftieth round would presumably depend on the plays on
      earlier rounds. In a tournament, each program played
      round-robin against every other program. (In addition,
      one program was played randomly.) Before deciding
      what overall strategy you would choose, think about the
      implications of each strategy.*

2. Suppose that you and another player are engaged in a game
   in which preplay communication is allowed.
   a. Suppose, further, that the game is changed so that you
      are prevented from using some of your strategies; can
      this possibly be to your advantage?
   b. Communications are cut off; can this be to your advan-
      tage? (Assume that you could make binding agreements
      when communication was available.)
   c. Suppose the rules in the original game required you and
      the other player to make simultaneous choices of strate-
      gy. However, the rules are changed so that your decision
      must be made first and the other player chooses after
      seeing what you have done; can this be to your advan-
      tage?

* This is a particularly important game and really worth some thought—take the
time to sort it out.

d. Will it be to your advantage or disadvantage if the other player learns what your utility function is?

3. A customer offers to buy a lamp from you, an antique dealer, for $5,000 within twenty-four hours. A wholesaler, who knows of the customer's offer, obtains the lamp for $3,000. At the last minute you get a note from the wholesaler offering the lamp for sale. You have only enough time to accept the offer and take your profit or turn it down and lose the sale. Would you accept the offer if:
   a. The selling price was (i) $4,000? (ii) $4,500? (iii) $4,950?
   b. Would it make any difference if you found out that the wholesaler mistakenly believed that the customer offered $7,000 rather than $5,000?

4. A speed limit was recently imposed in a community and the town council must now decide how strictly it should be enforced. By quantifying various costs and benefits to the community and the drivers—the time saved by speeding, the danger to the driver, the danger to the public in general, the penalties to the speeding driver, the cost of enforcement—the town game theorist arrives at the payoff matrix depicted in figure 5.2.

Figure 5.2

COMMUNITY

		ENFORCE THE LAW	DO NOT ENFORCE THE LAW
DRIVER	SPEED	(−190, −25)	(10, −5)
	DO NOT SPEED	(0, −20)	(0, 0)

The matrix reflects the payoffs if the law is enforced 100 percent of the time. Is it "cheaper" for the community to enforce or ignore the law? Is there a third alternative?

5. You and a stranger each pick a number from 1 to 10. A philanthropic organization pays the one who picked the *lower* number as many thousands of dollars as the number picked. (If you both pick the same number, the winner is determined by a coin toss.)

    a. What number should you choose if the game is played:
(i) once? (ii) fifty times?

    b. The game is to be played fifty times. On the first two
plays, the stranger chooses 10 and 9. What do you make
of this?

6. This last example is a kind of psychological test that measures your, and a hypothetical partner's, joint aspirations. Imagine that you and your partner are each given an algebraic
expression after you have been placed in separate rooms to
avoid communication. In example (a) that expression is $2X
+ 5Y - 100$. You choose a number X and your partner
chooses a number Y. If the numbers you choose jointly
make the expression positive, you get nothing; if they make
it negative or zero, you get $\$X$ and your partner gets $\$Y$. So
if you choose $X = 10$ and your partner chooses $Y = 2$, you
get $10 and your partner $2 since $(2)(10) + (5)(2) - 100 =
-70$, which is negative. You both get nothing if you choose
40 and your partner chooses 5, since $(2)(40) + (5)(5) - 100
= 5$, which is positive. You want to make X as large as
possible so long as it isn't so large as to reduce your payoff to
zero.

    To help visualize the problem, I have drawn the graph of
the equation $2X + 5Y - 100 = 0$ (and all the other graphs as
well). (See figure 5.3 *A–H*.) Interpreted geometrically, if the
point (X, Y) is below the graph or on it, you and your partner get $\$X$ and $\$Y$, respectively. If it is above the graph, you
both get nothing.

    If you are going to act intelligently, you should know
something about your partner; so assume, as usual, that he
or she is as bright as you are and has the same general
attitude toward money that you do.

THE theory of two-person, zero-sum games is unusual in
that it enables you to find solutions—solutions, moreover,
that are universally accepted. In this respect, zero-sum games
are unlike the actual problems that arise in everyday life,
which generally do not lend themselves to straightforward
answers. For non-zero-sum games we cannot do nearly as

Figure 5.3

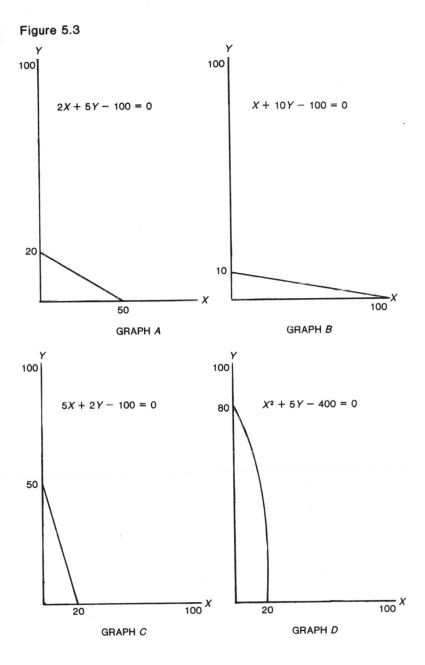

Figure 5.3 (*cont.*)

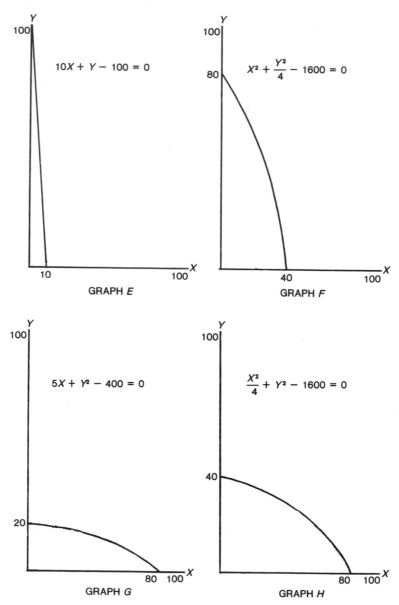

well. In most games of any complexity, there is no universally accepted solution; that is, there is no single strategy that is clearly preferable to the others, nor is there a single, clear-cut, predictable outcome. As a rule, we will have to be content with something less than the unequivocal solutions we obtained for zero-sum games.

For convenience, let's regard all two-person games as lying in a continuum, with the zero-sum games at one extreme. In a two-person game, there are generally both competitive and cooperative elements: the interests of the players are opposed in some respects and complementary in others. In the zero-sum game, the players have no common interests. In the completely cooperative game at the other extreme, the players have nothing but common interests. A pilot of an airplane and the operator in the control tower are engaged in a cooperative game in which they share a single common goal, a safe landing. Two sailboats maneuvering to avoid a collision, and two partners on a dance floor, are also playing a cooperative game. The problem in such a game is easy to solve, at least conceptually: it consists of coordinating the efforts of the two players efficiently (by means of dancing lessons, for instance).

The rest of the two-person games, and the ones with which this chapter will be primarily concerned, fall between these two extremes. Games with both cooperative and competitive elements are generally more complex, more interesting, and encountered more frequently in everyday life than pure competitive and cooperative games. Some examples of these games are: an automobile salesperson negotiating with a customer (both want to consummate the sale, but they differ on the price), two banks discussing a merger, two competing stores, and so forth. In each of these games the players have mixed motives. Also, there are many situations where the parties *seem* to have no common interests but really do.

Two nations at war may still honor a cease-fire, not use poison gas, and refrain from using nuclear weapons. In fact, zero-*sum* games are almost always an approximation to reality—an ideal that is never quite realized in practice. In chapter 4, for example, if the examples are varied even slightly, they cease to be zero-sum. In the marketing problem, it was assumed that total consumption was constant. But if, when the advertising schedule was changed, total consumption also changed (as in practice it would), the game would be non-zero-sum. And even if the assumption were valid but the firms conspired together to reduce the amount of advertising (and thus lower costs), the game would also be non-zero-sum.

Analyzing a Two-Person, Non-Zero-Sum Game

The easiest way to get some insight into the non-zero-sum game is to try to analyze one in the same way that the zero-sum game was analyzed. Suppose we start with the matrix shown in figure 5.4. Notice that in non-zero-sum games it is necessary to give the payoffs for *both* players, since the payoff of one player can no longer be deduced from the payoff of the other, as it can in zero-sum games. The first number in the parentheses is the payoff to player *I*, and the second number is the payoff to player *II*.

Figure 5.4

		PLAYER *II*	
		A	B
PLAYER *I*	a	(0, 0)	(10, 5)
	b	(5, 10)	(0, 0)

It is obvious what the players' common interests is to avoid the zero payoffs. But there still remains the problem of determining who will get 5 and who will get 10. One way to attack the problem, a way that was successful before, is to look at the game from the point of view of one of the players—say, player *I*. As player *I* sees it, the game has the matrix shown in figure 5.5. Since the payoffs to player *II* don't concern that player directly, they are omitted.

Figure 5.5

PLAYER *II*

		A	B
PLAYER *I*	a	0	10
	b	5	0

Ultimately, player *I* must decide what she wants from this game and how she should go about getting it. A good way to start is by determining what she can get without any help from player *II;* this, at any rate, is the *least* player *I* should be willing to settle for.

Using the techniques of the zero-sum game, player *I* will find that, by playing strategy *A* one-third of the time she can get an average return of 10/3. If player *II* plays strategy *A* two-thirds of the time, that's all player *I* can get.

If player *II* makes the same sort of calculation, using his own payoff matrix and ignoring the payoff to player *I,* he too can get 10/3 by playing strategy *A* one-third of the time. If player *I* plays strategy *a* two-thirds of the time, player *II* will not get any more.

So far, so good. It looks as if we are in much the same position as we were in the zero-sum game: each player can get 10/3; each player can be stopped from getting any more. Why not, then, call this the solution?

The trouble with this "solution" is that the payoffs are too low. If the players can manage to get together on either one of the non-zero payoffs, they will *both* do better. The argument used to support the (10/3, 10/3) payoff, which was sound enough in the zero-sum game, fails here. *Although each player can prevent the other from getting more than 10/3, there is no reason why he or she should.* It is no longer true that a player can get rich only by keeping an opponent poor. Although it is true that the players cannot be prevented from getting 10/3, they are foolish if they don't aspire to something better.

But how to go about doing better? Suppose player *I* anticipates everything that I have said so far and guesses that players *II* will play strategy *A* one-third of the time, to ensure a payoff of 10/3. If player *I* switches to the pure strategy *a*, player *II* will still get his 10/3, but player *I* will now get 20/3: twice as much as before. The trouble with this argument is that player *II* may get ambitious also. This player may also decide to double his return and, anticipating conservative play by player *I*, play the pure strategy *A*. If they both do this simultaneously, they will both get nothing. As in the zero-sum game, this sort of circular reasoning does not take you very far.

Up to now, my main purpose has been to give some idea of the kind of problems that may arise. To this end, and also for future reference, let us look at some examples.

The Gasoline Price War

Two competing gas stations buy gas at 20 cents a gallon and, together, sell a thousand gallons a day. They have each been selling gas at 25 cents a gallon and dividing the market evenly. One of the owners is thinking of undercutting the other. The prices, which are quoted in whole numbers, are set independently by each station in the morning and remain fixed for the day. There is no question of customer loyalty, so a lower price at either station attracts virtually all the sales. What price should each station owner set? If one owner knows the other is going out of business the next day, should that make a difference? What if the owner is going out of business ten days later? What if the owner is going out of business at some indefinite time in the future?

For each set of prices set by the gas stations, corresponding profits are entered in the matrix in figure 5.6. If gas station *I* charges 24 cents and gas station *II* charges 23 cents, gas station *II* captures the entire market; its owner makes 3 cents profit per gallon on a thousand gallons, or $30 profit. Gas station *I* makes no profit at all.

Figure 5.6

GAS STATION *II*'S PRICE PER GALLON

		25¢	24¢	23¢	22¢	21¢
	25¢	(25, 25)	(0, 40)	(0, 30)	(0, 20)	(0, 10)
	24¢	(40, 0)	(20, 20)	(0, 30)	(0, 20)	(0, 10)
GAS STATION *I*'S	23¢	(30, 0)	(30, 0)	(15, 15)	(0, 20)	(0, 10)
PRICE PER GALLON	22¢	(20, 0)	(20, 0)	(20, 0)	(10, 10)	(0, 10)
	21¢	(10, 0)	(10, 0)	(10, 0)	(10, 0)	(5, 5)

Let us start by trying to answer the easiest question: What price should a gas station set if its competitor is going out of business the next day? The first thing to note is that one should *never* charge less than 21 cents or more than 25 cents. When the price drops below 21 cents, there is no profit, and when it rises above 26 cents, you lose your market. Note also that you should never charge 25 cents, since this price is dominated by 24 cents. If your competitor charges either 24 or 25 cents, you do better by charging 24 cents, and if your competitor charges something else, you don't do any worse. A gas station should not charge 25 cents, then, and the owner may reasonably assume that a competitor won't either.

Once 25 cents is ruled out as a selling price for *either* gas station, you can, by the same reasoning, rule out 24 cents as a selling price, because it is dominated by a price of 23 cents. As a matter of fact, you can successively rule out every price but 21 cents.

There is a paradox here. Starting from the status quo, where the stations charged 25 cents a gallon and received $25 profit apiece, they managed, by reasoning logically, to get to a position where they charge 21 cents and receive only $5 profit apiece.

If one of the gas stations is known to be going out of business in ten days rather than in one day, it's a bit more complicated. The old argument is all right as far as it goes, but now we are concerned with the profits not only of the next day but of the following nine as well. And if a player drops his or her price one day, that player can be sure that the competitor will do the same on the next. It turns out, however, that even here an apparently airtight argument can be made for charging 21 cents. It is only when the competition continues indefinitely that the argument finally loses force.

This game is a variation of the prisoner's dilemma; I will say more about it later.

A Political Example

The state legislature is about to vote on two bills that authorize the construction of new roads in cities *A* and *B*. If the two cities join forces, they can muster enough political power to pass the bills, but neither city can do it alone. If a bill is passed, it will cost the taxpayers of both cities a million dollars and the city in which the roads are constructed will gain $10 million. The legislators vote on both bills simultaneously and secretly; each legislator must act on each bill without knowing what anyone else has done. How should the legislators from cities A and B vote?

The payoff matrix for this game is shown in figure 5.7. Since a city always supports its own bond issue, the legislators have only two strategies: supporting, or not supporting, the sister city.

Figure 5.7

		CITY B	
		SUPPORT A'S BOND ISSUE	WITHHOLD SUPPORT FROM A'S BOND ISSUE
CITY A	SUPPORT B'S BOND ISSUE	(8, 8)	(−1, 9)
	WITHHOLD SUPPORT FROM B'S BOND ISSUE	(9, −1)	(0, 0)

The entries in the matrix are in millions of dollars. As an example of how they were computed, suppose city A supports city B's bond issue, but B doesn't support A's. Then one bond issue passes, each city pays a million dollars, B gets $10 million, and A gets nothing. The net effect is that A loses a million dollars and B gets $9 million.

As before, the smartest thing seems to vote against the other city's bond issue. And, just as before, if both cities follow this strategy, they will both get nothing, rather than getting the $8 million apiece they could have gotten otherwise.

GAME THEORY

The Battle of the Sexes

Another interesting two-person, non-zero-sum game—the battle of the sexes—is described by R. Duncan Luce and Howard Raiffa (1957). A husband and his wife have decided that they will go either to the ballet or to the prize fights that evening. Both prefer to go together rather than going alone. While the husband would most prefer to go with his wife to the fights, he would prefer to go with her to the ballet rather than go to the fights alone. Similarly, the wife's first choice is that they go together to the ballet, but she too would prefer that they go to the fights together rather than go her own way alone. The matrix representing this game is shown in figure 5.8. The payoffs reflect the order of the players' preferences. This is essentially the same game I talked about at the opening of this chapter.

Figure 5.8

		HUSBAND	
		FIGHTS	BALLET
WIFE	FIGHTS	(2, 3)	(1, 1)
	BALLET	(1, 1)	(3, 2)

A Business Partnership

A builder and an architect are jointly offered a contract to design and construct a building. They are offered a single lump sum as payment and given a week to decide whether they will accept. The approval of both is required, so they must decide how the lump sum will be divided. The architect writes to the builder, suggesting that the profits be divided equally. The builder replies, committing himself to the contract in writing, provided he receives 60 percent of the profits. He informs the architect, moreover, that he is going on a two-week vacation, that he cannot be reached, and it is up to the architect to accept the contract on these terms or reject it entirely.

The architect feels she is being exploited. She feels that her services are as valuable as the builder's and that each should receive half the profits. On the other hand, the probable profit is large, and under different circumstances, she would consider 40 percent of such a total an acceptable fee. Should she accept the builder's offer? Note that the builder has no more options; only the architect has a choice—to accept or reject the contract. This is reflected in the payoff matrix shown in figure 5.9.

Figure 5.9

		BUILDER
ARCHITECT	ACCEPT OFFER	(40%, 60%)
	REJECT OFFER	(0, 0)

A Marketing Example

Each of two competing firms is about to buy an hour of television time to advertise its products. They can advertise either in the morning or in the evening. The television audience is divided into two groups. Forty percent of the people watch during the morning hours, and the remaining 60 percent watch in the evening; there is no overlap between the groups. If both companies advertise during the same period, each will sell to 30 percent of the audience watching and make no sales to the audience not watching. If they advertise during different periods, each will sell to 50 percent of the audience watching. When should they advertise? Should they consult before making a decision? The payoff matrix for the game is shown in figure 5.10.

Figure 5.10

		FIRM *II*	
		ADVERTISE IN MORNING	ADVERTISE IN EVENING
FIRM *I*	ADVERTISE IN MORNING	(12, 12)	(20, 30)
	ADVERTISE IN EVENING	(30, 20)	(18, 18)

The matrix entries represent the percentage of the total audience that each firm captures. If firm *I* picks an evening hour and firm *II* picks a morning hour, firm *I* sells to half of 60 percent, or 30 percent, and firm *II* sells to half of 40 percent, or 20 percent.

Some Complications

Two-person, zero-sum games come up in many different contexts, but they always have the same basic structure. By looking at the payoff matrix, you can pretty well tell "the whole story." This is not the case in non-zero-sum games. Besides the payoff matrix, there are many "rules of the game" that markedly affect the character of the game, and these rules must be spelled out before you can talk about the game intelligently. It is impossible to say very much on the basis of the payoff matrix alone.

To see this more clearly, go back to the first example discussed on page 82—in which the payoffs were (0, 0), (10, 5), and (5, 10). There were several details that I deliberately omitted but which are obviously important. Can the players consult beforehand and agree to their respective strategies in advance? Are these agreements binding? That is, will the referee or whoever enforces the rules insist that the agreement be carried out, or does the agreement have moral force only? Is it possible for the players, after one has received 5 and the other 10, to make payments to each other so they both receive 7 1/2? (In some games, they can; in others, they can't.) In the gasoline price war, is it feasible (legal) for the two businesses to conspire to fix prices?

It is to be expected that these (and other) factors will have a strong influence on the outcome of the game; but their actual effect is often different from what one would expect.

After you study the two-person, zero-sum game, certain aspects of the non-zero-sum game seem like something out of *Alice in Wonderland*. Many "obvious" truths—fixed stars in the firmament of the zero-sum game—are no longer valid. Look at a few:

1. You would think that the ability to communicate could never work to a player's disadvantage. After all, even if players have the option to communicate, they can refuse to exercise it and so place themselves in the same situation as if facilities for communication were nonexistent; or so it would seem. The facts are otherwise, however. The *inability* to communicate may well work to one player's advantage, and this advantage is lost if there is a way to communicate, even though no actual communication occurs. (In the zero-sum game, this doesn't come up. There the ability to communicate is neither an advantage nor a disadvantage, since the players have nothing to say to each other.)

2. Suppose that in a symmetric game—a game in which the payoff matrix looks exactly the same from both players' point- of view—player *I* selects a strategy first. Player *II* picks a strategy after seeing what player *I* has done. You would think that in a game where the players have identical roles except that player *II* has the benefit of some additional information, player *II*'s position would be at least as good as player *I*'s. In the zero-sum game, this situation could not

possibly be to player I's advantage. In the non-zero-sum game, however, it may. Let us carry the point a little further. It is often to a player's advantage to play before an opponent, even if the rules do not require that he or she do so, or, alternatively, to announce the chosen strategy so that his or her decision becomes irrevocable.

3. Suppose that the rules of a game are modified so that one of the players can no longer pick some of the strategies that were originally available to him or her. In the zero-sum game, the player might not lose anything, but he or she certainly could not gain. In the non-zero-sum game, the player might very well gain.

4. It often happens in a non-zero-sum game that a player gains if an opponent does not know his or her utility function. This is not surprising. What is surprising is that at times it is an advantage to have your opponent know your utility function, and he or she may be worse off after learning it. This doesn't come up in zero-sum games: there it is assumed that each player knows the other's utility function.

Communication

The extent to which players can communicate has a profound effect on the outcome of a game. There is a wide spectrum of possibilities here. At one extreme we have games in which there is no communication whatever between the players and the game is played only once (later I will discuss why this matters). At the other extreme are the games in which the players can communicate freely. Generally, the more cooperative the game—the more the players' interests coincide—the more significant is the ability to communicate. In the zero-sum game that is completely competitive, communication plays no role at all. In the completely cooperative game, the problem is solely one of communication, and so the ability to communicate is crucial.

Cooperative games in which the players can communicate freely present no conceptual difficulty. There may be technical difficulties, of course, such as exist when a control tower directs a pilot in heavy traffic. But two players who can't communicate directly, such as two sailboat captains trying to avoid a collision in choppy waters or two guerrillas behind enemy lines, have a problem.

In completely cooperative games, communication is an unalloyed blessing. In games in which the players have some conflicting interests, communication has a more complex role. To see this, look at the game shown in figure 5.11. There are two basic properties of this game that are worth noting. No matter which strategy a partner plays, *a player always does best if he or she plays strategy* B. Also, no matter what strategy a player finally selects, *he or she will always do better if the partner plays strategy* A. Putting these two together, we have a conflict: on the one hand, each player should play strategy *B;* on the other hand, each is hoping the other will play strategy *A.* It is to the advantage of both, moreover, that the conflict be resolved by both choosing *A* rather than *B.* How do the players go about getting this result?

Figure 5.11

PLAYER *II*

		A	B
PLAYER *I*	A	(5, 5)	(0, 6)
	B	(6, 0)	(1, 1)

If the game is played only once, the prospects for the (5, 5) outcome—sometimes called the cooperative outcome—are pretty bleak. Neither player can influence the partner's play, so the best each can do is play *B.* When the game is played repeatedly, it is a little different; it may be possible then to play in such a way that one's partner is induced to play A. And eventually it may be possible to get (5, 5) payoff.

Granted there is an opportunity to arrive at this payoff when the game is played repeatedly, how do you invite your partner to cooperate in a specific situation? One way to do it is unilaterally to adopt an apparently inferior strategy (such as strategy A in this illustration) and hope the other player catches on. If your partner turns out to be either stupid or stubborn, there is nothing to do but revert to the cutthroat strategy, B, and settle for a payoff of 1. Thus, in fact, it may be possible to communicate in a game even if there is no direct contact between the players.

This kind of tacit communication is generally ineffective, however, most likely because it is generally misunderstood. Merrill M. Flood (1952) described an experiment in which a game, similar to the gasoline price war discussed earlier, was played a hundred times by two sophisticated players. The players made written notes during the game, and these left little doubt that the players were not communicating. This same conclusion was also reached in other experiments, as will be seen later.

Even if the message requesting cooperation does get through, it may not be in the best interests of the receiver to accept it immediately. It has been suggested that the appropriate response is to "misunderstand" at first, and let the other player "teach" you; then, before the other player loses hope, join him or her in playing the cooperative strategy. Following this strategy, in the example you would make 6 while "learning" and 5 thereafter. There is some evidence that this is what actually happened in the Flood experiment. But this is a dangerous game, as shall be seen later in the discussion of computer play in prisoner's dilemma games.

An amusing way to exploit the inability of players to get together was suggested by Luce and Raiffa (1957). They would have a company offer each of two players two alternative strategies: a "safe" strategy and a "double-cross" one. If both play safe, they each receive a dollar; if both play double-cross, they each lose a nickel; and if one plays safe while the other plays double-cross, the safe player gets a dollar and the double-cross player gets a thousand dollars. The matrix in figure 5.12 shows the possibilities. Luce and Raiffa felt that as long as the players were prevented from communicating and only played the game once, the company would get some free advertising at no great expense.

Figure 5.12

		PLAYER *II*	
		SAFE	DOUBLE-CROSS
PLAYER *I*	SAFE	($1, $1)	($1, $1,000)
	DOUBLE-CROSS	($1,000, $1)	(−5¢, −5¢)

On the basis of the games we have seen up to now, the ability to communicate would seem to be an advantage. So far, the only messages contemplated are offers to cooperate, and these obviously must be in the interest of both players. Otherwise the offers would not be made or, if made, would not be accepted. But it is also possible to communicate threats. The game shown in figure 5.13 was devised by Luce and Raiffa. Compare what happens when the players can communicate with what happens when they can't.

If the players can't communicate, they obviously can't threaten. Player *I* can do no better than to play strategy *A*,

Figure 5.13

		PLAYER *II*	
		a	b
PLAYER *I*	A	(1, 2)	(3, 1)
	B	(0, −200)	(2, −300)

95

and player *II* can do no better than to play strategy *a*. But when the players are allowed to communicate, there is a radical change. Just what will happen is not altogether clear; to some extent, it will depend on circumstances that haven't been discussed. But, in any case, player *II*'s position deteriorates. If agreements made by the players can be enforced, player *I* can threaten to play strategy *B* unless player *II* commits herself to playing strategy *b*. If player *II* submits, player *I* will gain 2 and player *II* will lose 1, relative to the original (1, 2) payoff. If side payments are allowed, player *II*'s position becomes even worse. Not only can player *I* dictate player *II*'s strategy, but he can demand something under the table as well. True, player *II* can refuse to negotiate with player *I* and simply ignore his threats. It would still be in player *I*'s best interests to play strategy *A*, but who can say what would have a stronger influence on his behavior, his self-interest or his pique at being rebuffed? The point is that player *II* can avoid all these complications if communication is impossible.

The Order of Play

In the zero-sum game, the players pick their strategies simultaneously, neither player knowing the opponent's choice. If a player manages to find out the opponent's strategy in advance, he or she is far ahead and, in principle at least, the game is too trivial to be of interest. The non-zero-sum game is altogether different. The game may be far from trivial even when one player learns the opponent's strategy. What's more, the advantage of having this information may turn out to be a disadvantage. Let's look at an example.

A buyer and a seller are negotiating a contract in which the price per item and the quantity to be sold are still to be determined. According to established procedure, the seller first sets the price, which, once quoted, may not afterward be changed; and the buyer indicates the quantity he wishes to buy.

In this instance, a wholesaler can buy two items from the manufacturer, one for $4 and the other for $5. The retailer has two customers for these items, one of whom is willing to pay $9 and the other, $10. If the mechanism for negotiating is as has just been described, what strategies should the players adopt? What will be outcome be?

Certain obvious characteristics of this game should be noted. It is clearly to both players' advantage to get together and in some way share the potential $10 profit. Also, if they are to share this profit equally and obtain a "fair" outcome, the selling price should be set at $7.

The wholesaler may have her eye on something better than a fair outcome, however. If she sets the price at $8 rather than $7, it would still be in the retailer's best interests to buy both items, although he would make a profit of only $3 rather than $5. (If he buys only one item, his profit would be $2, and if he buys nothing, he would have no profit at all.) In effect, the negotiating mechanism that requires the wholesaler to make the first move allows her to put pressure on the retailer, to her own advantage.

True, the retailer need not act mechanically in his own "self-interest" and allow himself to be exploited. Moreover, in the pure bargaining game in which buyer and seller negotiate freely about quantity and price simultaneously, the players will not always arrive at a selling price of $7; personality factors may affect the price one way or the other. Even though the consequences of requiring the wholesaler to move first cannot be predicted precisely, however, the general effect is, clearly, to give the wholesaler the upper hand.

GAME THEORY

The Effect of Imperfect Information

The wholesaler-retailer game just described was one of a series of experiments Lawrence E. Fouraker and Sidney Siegel (1963) studied in the laboratory. In the series, a variation of the game was also played. The original buying and selling prices were kept the same, but the rules were changed somewhat and, with them, the character of the game. In the new version, the wholesaler knew only her own profits; the retailer knew both players' profits. In addition, the retailer knew that the wholesaler did not know.

The basic difference between the variation and the original game was the reaction of the retailer when the wholesaler quoted a high selling price. In the original game, in which both players were fully informed, a high price on the wholesaler's part was interpreted by the retailer as a sign of greed, and as a result he often refused to go along. But when the retailer knew that the wholesaler was unaware of how large a share of the total profit she was asking, the retailer generally accepted his fate philosophically and did the best he could under the circumstances. Thus the wholesaler often did better—and the retailer, worse—when she had less information.

A corollary of this is that it is often to a player's advantage to see that the partner is well informed. Suppose, in a labor–management dispute, that labor's demands are such that if granted, they would force the company out of business. In such an instance, the company should see to it that the union is informed of the effect of its demands. Of course, the company's interest is not so much to have the union know the truth for its own sake as it is to have the union believe that its goals are unattainable; and for this purpose, a lie will do just as well. The company, then, may try to deceive the union about, say, its inability to compete if it grants a raise. Or it may lie about its utility function, stating that it would prefer a prolonged strike to granting a raise, when in fact it would not. The union, for its part, might exaggerate the size of its strike fund. Ralph Cassady, Jr., (1957), describes some of the tactics used by competitors in a taxicab price war to "inform" (misinform) their opponents. They printed signs—which were never intended to be used—quoting rates considerably below the prevailing ones, and left them where they were sure to be seen by the competition. They also leaked the "information" that the company's owner had inherited a fortune—actually, he had inherited only a modest sum of money—and so gave the impression that the company had the capacity and the intention of fighting for some time.

In such a situation, players can gain if they can convince their opponent that they have certain attitudes or capabilities, whether they really have them or not. (If you are bargaining for an antique that you particularly want, it is a good idea not to let the seller know it.) If players really have these capabilities or attitudes, we have the situation spoken of earlier, where a player gains when the partner is better informed.

The Effect of Restricting Alternatives

Players are sometimes prevented from using some of their strategies. It is one of the paradoxes of non-zero-sum games that this restriction of a player's choice may be turned to that player's advantage. On the face of it, this seems absurd. How is it possible for a player to gain if prevented from using certain strategies? If it were to the player's advantage to avoid using certain strategies, could he or she get the same effect simply by not using them? No, the player could not; pretending you don't have certain alternatives is not the same as not having them. Suppose, for example, in the labor–

management dispute, that there were strict wartime controls in effect that prevented the company from raising wages. In that case, a union that under different circumstances might go on strike would very likely continue working, without objection. By the same token, it would be to the woman's advantage, in the "battle of the sexes" example, if she tended to faint at the sight of blood and could not, therefore, sit through a prize fight.

When circumstances do not restrict players' alternatives, they may try to gain an advantage by unilaterally restricting them themselves. This is not always so effective, however. If the woman in the "battle of the sexes" commits herself and her husband to the ballet by buying two tickets in advance, her husband may refuse to accompany her out of pique. If she cannot attend a fight because she faints at the sight of blood, a factor over which she has no control, he may view it differently.

This principle of limiting your alternatives to strengthen your position can be applied in other ways. We have already seen one application in the "business partnership" example. Another application is the hypothetical weapon called the doomsday machine. A weapon of great destructive power, it is set to go off *automatically* whenever the nation that designed it is attacked.

The point of building a doomsday machine is this: so long as the defending nation preserves its option to withhold retaliation, the way is open to an unpunished attack. A potential aggressor may be tempted to attack and then, by threatening an even worse attack, inhibit the defending nation from retaliating. With the doomsday machine, the defending nation forecloses one of its options; it cannot help but retaliate. Here again, it is to a player's advantage to keep the partner well informed; if you have a doomsday machine, it is a good idea to let everyone know it.

Threats

A *threat* is a statement that you will act in a certain way under certain conditions. It is like the doomsday machine in that it limits your future actions: "If you cut your price by five cents, I'll cut mine by a dime." But it is different because this self-imposed restriction isn't binding; you can always change your mind. The purpose of a threat is to change someone's behavior: to make that person do something he or she would not do otherwise. If the threat is carried out, it will presumably be to the detriment of the party that is threatened, but often it is also to the disadvantage of the party making the threat.

A threat is effective only to the extent that it is plausible. The greater the price the party making the threat must pay to carry it out, the less plausible the threat. This leads to the following paradox: if the penalty to the people making the threat is very high, they will be reluctant to burn their bridges behind them by committing themselves irrevocably to the threat. But it is precisely the threatener's failure to burn the bridges that is the greatest inducement to ignoring the threat. When you are bargaining for a new car, you can discount such statements as "I won't let it go for a penny under $2,000." If the seller is convinced that at this price no sale will be made, the seller may later choose to ignore his or her own threat. In a store where prices are fixed by the management, however, and the final decision is made by a disinterested salesperson on salary, the "threat" of not lowering prices is essentially irrevocable. Of course, in that case the store must accept the possibility that it will occasionally lose a sale. In the "battle of the sexes," one of the players may threaten to go it alone, but, again, the threat is not irrevocable, and the other player may threaten to resist. In the wholesaler-retailer example, the wholesaler was committed to her quoted selling price by the rules of the game, and, as a result, the threat was much stronger.

Often both players are in a position to threaten. In a bargaining game, for example, both the buyer and the seller can refuse to complete the sale unless the price is right. In the battle of the sexes, each player could threaten to attend his or her preferred entertainment alone. Occasionally, however, it happens that only one of the players is in a position to threaten. Such a situation was cleverly exploited by Michael Maschler (1963) in his analysis of an inspection model designed to detect illicit nuclear testing.

The two players represented in this model are countries that have signed an agreement outlawing nuclear testing. One of them is considering violating the treaty; the other wants to be able to detect a violation if it occurs. (In reality, a country may play either role, or both at the same time.) The inspecting country has the benefit of detecting devices that indicate natural as well as artificial disturbances; it is allotted a quota of on-site inspections. The strictly mathematical problem consists of timing the inspections so as to maximize the probability of discovering a violation if there is one and determining when, if ever, the potential treaty violator would test a nuclear device.

This, obviously, is not a zero-sum game, since presumably both the inspector and the inspected would prefer that there be no violations to having violations that are subsequently discovered. In any case, this is assumed in the model. Maschler showed that the inspector actually does best by announcing a strategy in advance and keeping to it, much as the wholesaler gained by setting a high price. (This is based on the assumption that the violator believes the announcement and then acts according to his or her own self-interest. There is no reason why the violator shouldn't believe it, for it is in the inspector's self-interest to tell the truth.) Why can't the tester use similar tactics; that is, announce an intention to cheat in some set manner and have the inspector make the best of it? If you look at the bare payoff matrix, it is clear that the tester can, but the political realities are such that this is not feasible.

Henry Hamburger (1979) tells how a community might use this same principle to enforce its speeding laws. He estimated the utility to the potential speeder and community of each of a number of factors: (1) the time saved by speeding; (2) the risk to the driver; (3) the driver's fine when caught; (4) the cost of enforcement, and (5) the hazard to others of speeding—and arrives at the matrix shown in figure 5.14.

Figure 5.14

		COMMUNITY	
		ENFORCE	IGNORE
DRIVER	VIOLATE	(−190, −25)	(10, −5)
	COMPLY	(0, −20)	(0, 0)

It appears that whatever the driver does, the community does better to ignore the speeding; in such a case the driver can be expected to ignore the laws and the community's payoff will be minus 5. But if the community announces and implements a policy of enforcing the law 10 percent of the time, the situation can be depicted in figure 5.15.

Figure 5.15

		COMMUNITY
DRIVER	VIOLATE	(−10, −7)
	COMPLY	(0, −2)

Now it is to the driver's advantage to comply and the community's loss is only 2.

Once again, looking only at figure 5.14, the driver could

103

also adopt a strategy in advance to force the community's strategy; however, in the real world this is not a practical alternative.

Binding Agreements and Side Payments

When players negotiate, they often reach some sort of agreement. In some games there is no mechanism for enforcing agreements and the players can break their word with impunity. But in other games this is not the case. Though any agreement that is reached is reached voluntarily, once made, it is enforced by the rules. This possibility of making *binding agreements* has a strong influence on the character of the game.

Let us look at an anecdote related by Merrill M. Flood (1952). Though this "game" actually had more than two players, it is pertinent here.

Flood wanted one of his children to baby-sit the others. He suggested that a fair way of selecting the sitter and setting a price for the service was to have the older children bid against one another in a backward auction. That is, he would start with a price of $4—the most he was willing to pay—and the children would bid successively, each bid lower than the last one, until the bidding stopped. The last person to bid would baby-sit for the agreed price.

It wasn't long before the children hit upon the possibility of collusion. When they asked about it, their father said he would allow them to "rig" the bidding if they satisfied two conditions: the final price must be no more than the ceiling of $4 he had set originally, and the children must agree among themselves in advance who would do the baby-sitting and how the money would be divided. As it happened, the children did not reach an agreement. Several days later, a bona fide auction was held and the final price was set at 90 cents. Thus the mere opportunity to communicate and the existence of a mechanism to enforce agreements are not sufficient to guarantee that an agreement will be reached. In this game the failure to get together resulted in an outcome markedly inferior to what might have been if the players had acted together.

In some games it is possible for one player to affect the actions of another by offering a "side payment," a payment made "under the table." This was the case in Flood's baby-sitting example. If an agreement had been reached, the child who baby-sat would have paid the other children a certain amount in return for staying out of the bidding.

In many games, however, the players cannot or may not make side payments. Sometimes this is a matter of policy—when the government invites private companies to bid on a contract, it responds to a cooperative strategy on the bidders' part (and to the corresponding side payments) with less than enthusiasm. Sometimes side payments are impractical because there is no unit that can be transferred from one player to the other. In the "battle of the sexes," the pleasure the wife feels when her husband takes her to the ballet is simply not transferable. (She may be able to reciprocate in a future game, however; perhaps next time she'll go to the fights.) Similarly, one legislator is barred from giving another any direct payment such as money in return for support, but the legislator may repay the supporter in kind in the future.

The simple example in figure 5.16 will help clarify the role played by side payments. If it is impossible to make side payments, player *II* can do no better than choose strategy *B* and take a dollar. But if side payments are possible (and if the players can make binding agreements), it is a very different game. Player *II* is then in a position to demand a sizable portion of player *I*'s thousand dollars, and if player *I* refuses, he may be left with only $100. Whether player *II* will actually follow through on her threat and sacrifice a dollar, player *I* must decide for himself.

Figure 5.16

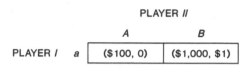

You cannot predict with any degree of assurance what will happen in this kind of game, or even to prescribe what should happen in theory. In practice, what happens no doubt depends strongly on the prize at stake. In games 1 and 2 in figure 5.17, the players may communicate and may make binding agreements, but side payments are not permitted.

Figure 5.17

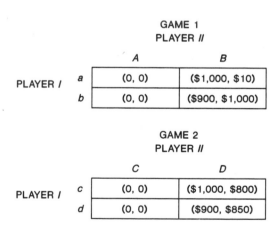

In game 1, player *II* will very likely try to get $1,000 by playing strategy *B* and persuading player *I* to play strategy *b*. Player *II*'s threat is to play strategy *A*, which will reduce player *I*'s payoff to nothing, at a cost of only $10 to himself.

Prudence seemingly dictates that player *I* should settle for the $900.

In game 2, the discussion might go much the same way, but now player *II*'s threat is less plausible. Her "spite" strategy, strategy *C*, is a lot more expensive, though her threat to player *I* is essentially the same. Still, it is difficult to say what the players will do. If player *I* does decide to call player *II*'s bluff, however, he plays a more dangerous game in 1 than in 2.

An Application to a Prison Break

A vivid illustration of some of these factors at work in real life was detailed in a news article in the *New York Times* on June 28, 1965. The article described a prison disturbance in which two guards were taken hostage. The warden refused to negotiate with the prisoners as long as the guards were held captive, and the guards were eventually freed, unharmed. The warden was quoted as saying: "They wanted to make a deal. I refuse to make deals at any time. Therefore I didn't listen, and I don't know what the deal was about." Figure 5.18 analyzes the "game."

Figure 5.18

		THE WARDEN	
		NEGOTIATE	DO NOT NEGOTIATE
PRISONERS	HARM GUARDS	*A*	*C*
	DO NOT HARM GUARDS	*B*	*D*

We can start by ruling out *A*, since the prisoners gain nothing by harming the guards if they are released. Of the three remaining possibilities, the prisoners most prefer *B*, then D, and least of all, C, which would mean additional punishment without any compensating gain. The warden wanted D the most, then B (presumably), and finally C.

The only chance of negotiating the prisoners had was to threaten to harm the guards unless they were released and to hope that the threat would be believed. But the warden simply cut communications; he refused to hear "what the deal was about." In effect, he committed himself to a strategy, not freeing the prisoners, and forced the prisoners to make a choice between C and D. His hope was that they would make the best of a bad situation by choosing D rather than C. And so they did—but if they had been sufficiently vindictive, they might have acted otherwise. Whether the warden made the proper choice is a question about which reasonable people might differ. Indeed, the warden and the hostages probably did.

The Prisoner's Dilemma

Two men suspected of committing a crime together are arrested and placed in separate cells by the police. Each suspect may either confess or remain silent, and each one knows the possible consequences of his action. These are: (1) If one suspect confesses and his partner does not, the one who confessed turns state's evidence and goes free and the other one goes to jail for twenty years. (2) If both suspects confess, they both go to jail for five years. (3) If both suspects remain silent, they both go to jail for a year for carrying concealed weapons—a lesser charge. We will suppose that there is no "honor among thieves" and each suspect's sole concern is his own self-interest. Under these conditions, what should the criminals do? The game, shown in figure 5.19, is the celebrated prisoner's dilemma which was originally formulated by A. W. Tucker and which has become one of the classical problems in the short history of game theory.

Figure 5.19

		SUSPECT *II*	
		CONFESS	DO NOT CONFESS
SUSPECT *I*	CONFESS	(5 yrs., 5 yrs.)	(go free, 20 yrs.)
	DO NOT CONFESS	(20 yrs., go free)	(1 yr., 1 yr.)

Let us look at the prisoner's dilemma from the point of view of one of the suspects. Since he must make his decision without knowing what his partner will do, he must consider each of his partner's alternatives and anticipate the effect of each of them on himself.

Suppose his partner confesses; our man must either remain silent and go to jail for twenty years, or confess and go to jail for five. Or, if his partner remains silent, he can serve a year by being silent also, or win his freedom by confessing. Seemingly, in either case, *he is better off confessing!* What, then, is the problem?

The paradox lies in this. Two naïve prisoners, too ignorant to follow this compelling argument, are both silent and go to prison for only a year. Two sophisticated prisoners, primed with the very best game-theory advice, confess and are given five years in prison in which to contemplate their cleverness.

We will return to this argument in a moment, but before we do, let's consider the essential elements that characterize this game. Each player has two basic choices: he can act "cooperatively" or "uncooperatively." When all the players act cooperatively, each does better than when all of them act uncooperatively. For any fixed strategy (ies) of the other player(s), a player always does better by playing uncooperatively than by playing cooperatively.

GAME THEORY

In the following examples, taken from many different contexts, these same basic elements appear.

1. Two different firms sell the same product in a certain market. Neither the product's selling price nor the total combined sales of both companies vary from year to year. What does vary is the portion of the market that each firm captures, and this depends on the size of their respective advertising budgets. For the sake of simplicity, suppose each firm has only two choices: spending $6 million or $10 million. The size of the advertising budget determines the share of the market and, ultimately, the profits of each company, as follows:

If both companies spend $6 million, they each get a $5 million profit. If a company spends $10 million when its competitor spends only $6 million, its profit goes up to $8 million at the expense of its competitor, who now loses $2 million. And if both companies spend $10 million, the extra marketing effort is wasted, since the market is fixed and the relative market position of each company remains the same; consequently, the profit of each company drops to $1 million. No collusion is allowed between the firms. The game is shown in figure 5.20.

Figure 5.20

		CORPORATION II	
		SPEND $6 MILLION	SPEND $10 MILLION
CORPORATION I	SPEND $6 MILLION	($5 million, $5 million)	(−$2 million, $8 million)
	SPEND $10 MILLION	($8 million, −$2 million)	($1 million, $1 million)

2. There is a water shortage and citizens are urged to cut down on water consumption. If each citizen responds to these requests by considering his or her own self-interest, no

one will conserve water. Obviously, any saving by an individual has only a negligible effect on the city's water supply, yet the inconvenience involved is very real. On the other hand, if everyone acts in his or her own self-interest, the results will be catastrophic for everyone.

3. If no one paid taxes, the machinery of government would break down. Presumably all citizens would prefer that everyone, including themselves, pay taxes to having no one pay taxes. Better yet, of course, everyone would pay taxes except the individual him- or herself.

4. After several years of overproduction, farmers agree to limit their output voluntarily in order to keep the prices up. But no one farmer produces enough seriously to affect the price, so each starts producing what he or she can and selling it for what it will bring, and once again there is overproduction.

5. Two unfriendly nations are preparing their military budgets. Each nation wants to obtain a military advantage over the other by building a more powerful army, and each spends accordingly. They wind up having the same relative strength, and a good deal poorer.

As we can see from these examples, this kind of problem comes up all the time. For convenience's sake, let us fix our attention on a single game, using the first example, that of two companies setting their advertising budgets. The game will be the same, except for an additional condition. The budget will not be set once but, more realistically, we will assume that it is fixed annually for a certain period of time; say, twenty years. When each company decides on its budgets for any given year, it knows what its competitor spent in the past.

In discussing the prisoner's dilemma, we had decided that if the game is played only once, the prisoners have no choice but to confess. The same line of reasoning applied here leads to the conclusion that the firms should each spend $10 million. But when the game is played repeatedly, the argument loses some of its force. It is still true that if you spend $10 million in a given year, you will always do better than if you spend $6 million *that year*. But if you spend $10 million in one year, you are very likely to induce your competitor to spend $10 million the next year, and that is something you don't want. A more optimistic strategy is to signal your intent to cooperate by spending $6 million and hope your competitor draws the proper inference and does the same. This strategy could lead to a cooperative outcome and in practice often does. But in theory there is a problem.

The argument that spending $6 million in any one year tends to encourage your competitor to spend the same amount the next year is all very well for the first nineteen years, but it clearly breaks down in the twentieth year. In the twentieth year, there is *no next year*. When the firms reach the twentieth year, they are in effect in the same position they are in when they play the game only once. If the firms want to maximize their profits, and we assume that they do, the argument favoring the uncooperative strategy is as compelling now as it was then.

But the argument doesn't end there. Once the futility of cooperating in the twentieth year is recognized, it follows that there is no point in cooperating in the nineteenth year either. And if there is no chance of inducing a cooperative response in the nineteenth year, why cooperate in the eighteenth year? Once you fall into the trap, you fall all the way: there is no point in cooperating in the eighteenth, the seventeenth . . . or the first year either. If you accept the argument supporting the uncooperative strategy in the single instance, it follows that you must play uncooperatively not only in the last of a series but in every individual trial as well.

It is when the prisoner's dilemma is played repeatedly—

and not for a fixed number of trials but for an indefinite period—that the cooperative strategy comes into its own. And these are precisely the conditions under which the prisoner's dilemma is often played. Two competing firms know that they won't be in business forever, but they generally have no way of knowing when death, merger, bankruptcy, or some other force will end their competition. Thus the players can't analyze what will happen in the last "play" and then work backward from there, for nobody knows when the last "play" will be. The compelling argument in favor of the uncooperative strategy breaks down, then, and we breathe a sigh of relief.

This is really the point. The prisoner's dilemma has one characteristic that makes it different from the other games that have been discussed. As a rule, when analyzing a game, you are content if you can say what rational players should do and predict what the outcome will be. But in the prisoner's dilemma the uncooperative strategy is so unpalatable that the question most people try to answer is not What strategy should a rational person choose? but How can we justify playing a cooperative strategy? Many different answers to this last question have been proposed. Let us consider a few.

The Prisoner's Dilemma in the Past

The prisoner's dilemma is important because it presents in capsule form a problem that arises in many different ways. Some of the manifestations of the prisoner's dilemma were discussed long before there was a theory of games. Thomas Hobbes, the political philosopher, examined one version of the dilemma in which the "players" were the members of society.

Society, Hobbes conjectured, was originally in a state of anarchy. Constant warfare and banditry were the consequence of each individual's trying to further his own narrow self-interest; it was a society in which *B* might murder *C* for a shiny bauble, and *B* might be murdered in turn for the same reason. It would be to everyone's individual advantage, Hobbes felt, if restrictions were imposed and enforced; that is, *B* would prefer to give up the chance of getting the bauble in return for additional security. Hobbes saw the social contract as an enforced cooperative outcome. In *The Leviathan* (1962) he described the creation of a government (preferably a monarchy) "as if every man should say to every man, I authorize and give up my right of governing myself, to this man or to this assembly of men, on this condition that thou give up thy right to him, and authorize all his actions in a like manner" (p. 104). (Whether historically this is an accurate picture is unimportant; what matters is how the problem was perceived and how it was resolved.)

Hobbes, after pointing out the disadvantages of the uncooperative outcome, suggests that the many independent decisions to cooperate or not be taken out of the hands of the people that make up society. In effect, society should submit to compulsory arbitration, and the government should play the role of the arbiter. This is not an uncommon point of view. Luce and Raiffa, in *Games and Decisions* (1957), make the same point: "Some hold the view that one essential rule of government is to declare that the rules of social 'games' be changed whenever it is inherent in the game situation that the players, in pursuing their own ends, will be forced into a socially undesirable position" (p. 97).

On a more modest scale, the noted sociologist George Simmel (1955) recognized that competing businesses are often faced with what amounts to a prisoner's dilemma. He described the behavior of businessmen who, in effect, are playing this game for an indefinite number of times:

> Inter-individual restriction of competitive means occurs when a number of competitors voluntarily agree to renounce certain practices of outdoing one another—whereby the renunciation by one of them is valid only so long as the other too observes it. Examples are the arrangement among booksellers of a given community to extend none or more than five or ten percent discount; or the agreement among store owners to close their businesses at eight or nine o'clock, etc. It is evident that here mere egoistic utility is decisive: the one does without those means of gaining customers because he knows that the other would at once imitate him if he did not, and that the greater gain they would have to share would not add up to the greater expenses they would likewise have to share. . . . In Economics, the third party is the consumer; and thus it is clear how the road toward cartelization is taken. Once it is understood that one can do without many competitive practices provided the competitor does likewise, the result may not only be an even more intense and purer competition, which has already been emphasized, but also the opposite. The agreement may be pushed to the point of abolishing competition itself and of organizing enterprises which no longer fight for the market but supply it according to a common plan. . . . This teleology, as it were, transcends the parties, allowing each of them to find its advantage and achieves the seeming paradox that each of them makes the opponent's advantage its own. (p. 76)

Note that Simmel looks at the consumer as an outsider, who, though affected by what the competing businesses do, has no control over what happens. In effect, the consumer is not a player. While cooperation between businesses is easily turned to their mutual advantage, the effects can easily be antisocial as far as society as a whole is concerned. And so society prohibits "cooperative play" in the form of trusts, cartels, price fixing, and bribery.

115

This theme—the tendency of competitors to avoid mutually destructive price competition—is also considered by John Kenneth Galbraith. In *American Capitalism: The Concept of Countervailing Power* (1952), he says: "The convention against price competition is inevitable. . . . The alternative is self-destruction" (p. 112). Price competition, then, tends to be replaced by competition in sales and advertising. When there are only a few sellers, interindustrial competition diminishes. If a few large industries are bargaining with their workers on wages, there ceases to be a wage competition among the individual industries to attract workers. There is, rather, bargaining in between the "countervailing powers" of labor and management.

Problems of the prisoner's dilemma type, in one form or another, have been around for some time. The cooperative strategy is generally accepted as the "proper" one (except when its effect is antisocial), sometimes for ethical reasons. Immanuel Kant (1959) asserted that a person should decide if an act is moral by examining the effect of everyone's acting similarly; the golden rule says much the same thing. More recently Anatol Rapoport in *Fights, Games and Debates* (1960) asserted that there are considerations other than a player's narrow self-interest that he or she should take into account when choosing play. If there is to be any hope of reaching the elusive cooperative outcome, Rapoport feels, it is necessary that the players accept certain social values, and having accepted these values, players should cooperate even in the one-trial, version of the prisoner's dilemma. His argument goes like this:

Each player presumably examines the whole payoff matrix. The first question he asks is "When are we both best off?" The answer in our case is unique: at [the cooperative outcome]. Next "What is necessary to come to this choice?" Answer: the assumption that whatever I do, the other will, made by both parties. The conclusion is, "I am one of the parties; therefore I will make this assumption." (p. 177)

Rapoport is well aware that his point of view is in conflict with "'rational' strategic principles" of self-interest; he simply rejects these principles. The minimax strategy in the two-person, zero-sum game, he asserts, is also based on an assumption: that one's opponent will act rationally; that is, in accordance with his or her own self-interest. If in this game one's opponent fails to act rationally, the minimax strategy will fail to exploit the opponent's errors. Just as you may be mistaken when you assume your opponent will be rational in the zero-sum game, you may be mistaken in the non-zero-sum game when you anticipate your partner's goodwill.

Since most people are generally reluctant to accept the uncooperative strategy as the proper one, there is a temptation, whenever a way out is proposed, to take it. Despite my sympathy with Rapoport's effort, however, I do not believe the paradox of the prisoner's dilemma has really been overcome. To see why, let us go back a bit.

In the prisoner's dilemma examples discussed in this chapter, the payoffs were stated in terms of "years in prison" or net profits, rather than utiles. The actual utilities involved are only suggested by such statements as: "Each player is only concerned with his own self-interest," or "Each company only wants to maximize its own profits." For the sake of simplicity, a more formal description in terms of utilities was avoided. But at bottom are certain assumptions that, though stated only inexactly, are critical. If they are not valid, we may be playing a radically different game from the one we think we are playing. If, in the original prisoner's dilemma, a prisoner would prefer to spend a year in jail along with his partner rather than go free, knowing his partner was serving twenty years, the argument for confessing breaks down. But then the game could hardly be called a prisoner's dilemma. The original paradox has in fact not been resolved. This is the objection to Rapoport's argument: his golden rule approach assumes the problem out of existence. If players are as concerned with a partner's payoff as they are with their own, the game isn't a prisoner's dilemma; and if each player is interested solely in his or her own payoff, Rapoport's comments aren't pertinent.

The analogy between the assumptions that are made in the zero-sum and in the non-zero-sum games is not very convincing either. In the zero-sum game, you can get the value of the game whether your opponent is good, bad, or indifferent; you do not have to assume that he or she is rational. Rapoport states as much but goes on to say that by playing minimax, you lose the opportunity of exploiting errors, and with a foolish opponent you should not be content to get only the value of the game.

This is not entirely true. There are games—we have seen some of them already—in which an inferior strategy on one player's part and the minimax strategy on the other's will lead to a payoff that is greater than the value of the game for the minimax player. But even in those games in which playing minimax precludes your getting more than the value of the game, the analogy is questionable. In order to exploit your opponent's weakness, it is not enough to know he or she will deviate from the minimax; you must also know how. Suppose, for example, you are going to match pennies *once* with a simpleton and you decide to assume that she will *not* play rationally. Specifically, you believe that she is quite capable of playing one side of the coin with a probability greater than one-half. But which side? And how can you exploit her inferior strategy if you don't know which side?

As a rule, when you play a minimax strategy in a zero-sum game it is not because you have faith in your opponent's rationality but because you have no other, more attractive alternative—and this is true even when you suspect your opponent is capable of making a mistake.

In the prisoner's dilemma, however, the assumption that your partner will cooperate is really an assumption. Unless you are a masochist, if you play cooperatively you must believe as an act of faith that your partner will too. Even if your partner cooperates and seemingly justifies your act of faith, some players will still question your choice since you could have done better still by playing uncooperatively. This attitude may seem piggish, but, then, players don't look to game theorists for moral principles; they already have their own. All they ask is to find a strategy that will suit their purpose, selfish or otherwise.

The difference between the assumptions made in the zero-sum and in the non-zero-sum games is even more clear when they fail. And there is no question that people often *do* fail to play cooperatively in society's prisoner's dilemmas. In non-zero-sum games, cooperating with a partner who doesn't cooperate with you leads to disaster; in zero-sum games, the worst that can happen when you play minimax is that you lose an opportunity to swindle your opponent.

The Nash Arbitration Scheme

Players in a bargaining game are in an awkward position. They want to make the most favorable agreement that they can, while avoiding the risk of making no agreement at all; and, to a certain extent, these goals are contradictory. If one party indicates a willingness to settle for any terms, even if the gain is only marginal, he or she will very likely arrive at an agreement, but not a very attractive one. On the other hand, if he takes a hard position and sticks to it, he is likely to reach a favorable agreement if he reaches any agreement at all—but he stands a good chance of being left out in the cold. A car dealer who is eager to sell a car will hide this fact from a buyer; but she will go to great lengths to determine how much she must drop her price to make a sale—even to the extent of using hidden microphones.

Even when a player is reconciled to receiving only a modest gain and pushes hard for an agreement, his or her eagerness is often interpreted as weakness, leading the partner to stiffen his or her demands and actually lessening the chance of an agreement. This is what often happens when one of two warring nations sues for peace. In a labor–management dispute that actually took place, one party who had a weak heart offered favorable terms in order to reach an immediate settlement. What happened was very different from what had been anticipated. Instead of waiving the usual bargaining procedure, the other party became suspicious and then resistant. The terms of the ultimate settlement were identical to those originally offered, but they were arrived at only after very hard bargaining.

One way of circumventing the actual bargaining process, at least in principle, is to have the terms of the agreement set by arbitration. In this way you can avoid the danger of not reaching a settlement at all. The problem is to establish arbitration that somehow reflects the strengths of the players in a realistic way, so that you get the effects of negotiation without risk. John Nash (1950) suggests the following procedure.

He begins by assuming that two parties are in the process of negotiating a contract. They might be management and labor, two countries formulating a trade agreement, a buyer and seller, and so forth. For convenience, and with no loss of generality, he assumes that a failure to agree—no trade, no sale, a strike, and so on—would have a utility of zero to both players. Nash then selects a single arbitrated outcome from all the agreements that the players have it in their power to make: that outcome in which the product of the players' utilities is maximized.* This scheme has four desirable properties that he feels justify its use, and it is the *only* one that does. The four properties are:

1. *The arbitrated outcome should be independent of the utility function.* Any arbitrated outcome should clearly depend on the preferences of the players, and these preferences are expressed by a utility function. But, as we saw earlier, there are many utility functions to choose from. Since the choice of utility function is completely arbitrary, it is reasonable to demand that the arbitrated outcome not depend on the utility function selected.

2. *The arbitrated outcome should be Pareto optimal.* Nash considered it desirable that the arbitrated outcome be Pareto optimal; that is, that there should not be any other outcome in which both players simultaneously do better.

3. *The arbitrated outcome should be independent of irrelevant alternatives.* Suppose there are two games A and B in which every outcome of A is also an outcome of B. If the arbitrated outcome of B turns out also to be an outcome of A, this outcome must also be the arbitrated outcome of A. Put another way, the arbitrated outcome in a game remains the arbitrated outcome even when other outcomes are eliminated as possible agreements.

4. *In a symmetric game, the arbitrated outcome has the same utility for both players.* Suppose the players in the bargaining game have symmetric roles. That is, if there is an outcome that has a utility of x for one player and a utility of y for the other, there must also exist an outcome that has a

---

* It should be noted that players may not only obtain utility pairs associated with simple agreements but may also obtain intermediate pairs by using coordinated strategies as well. Suppose, for example, that in the battle of the sexes, going to the ballet has a utility of 4 for the husband and 8 for the wife, and going to the fights has a utility of 6 for the husband and 2 for the wife. Each may obtain a utility of 5 if they let a toss of a coin determine the choice of the evening's entertainment.

utility of *y* for the first player and *x* for the second player. In such a game, the arbitrated outcome should have the same utility for both players.

Before the Nash arbitration scheme can be applied, the utility function of both players must be known. This is its biggest disadvantage, for not only are the utilities not always known, they are often deliberately obscured by the players. If a player's utility function is misrepresented, it can be turned to his or her advantage. This is reassuring in a way, for it suggests the scheme is realistic. In real life also the utility function is often misrepresented, as we have seen.

It is important to realize that the Nash scheme is neither enforceable nor a prediction of what will happen. It is, rather, an a priori agreement obtained by abstracting away many relevant factors such as the bargaining strengths, of the players, cultural norms, and so on. (In this respect it is similar to the Shapley value that will be discussed later.) As a matter of fact, the Nash outcome often appears to be unfair: it tends to make the poor poorer and the rich richer. This is to be expected, however. A rich player is often in a stronger position than a poor one. To see how this works, consider the following example.

Suppose a rich woman and a poor woman can get a million dollars if they can agree on how to share it between them; if they fail to agree, they get nothing. In such a case, the Nash arbitration scheme would generally give the rich woman a larger portion than it would give the poor woman, because of the difference in their utility functions. Let's take a moment to see why.

When *relatively* large amounts of money are involved— that is, amounts of money that are large relative to what a person already possesses—people tend to play it safe. Most people, unless they are very rich, would prefer a sure million dollars to an even chance of getting $10 million, although they would prefer an even chance of getting $10 to a sure dollar. But a large insurance company would prefer the even chance of $10 million, and in fact gladly accepts much less attractive risks every day. This indifference to the difference between large sums of money is reflected in the poor woman's utility function much more strongly than in the rich woman's. The relative attractiveness of $1 and $10 to the poor woman would be like the relative attractiveness of $1 million and $10 million to a woman who is very wealthy. A utility function that correctly reflects the poor woman's situation would be the square-root function: $100 would be ten utiles, $1 would be one utile, $16 would be four utiles, and so on. Thus the poor woman would be indifferent to an even chance at $10,000 and a sure $2,500. (The specific choice of the square-root function is of course arbitrary; many others would do as well.) It may be assumed that the rich woman's utility function is identical to the money in dollars. Under these conditions, Nash's suggested outcome would be that the rich woman gets two-thirds of the million dollars and the poor woman only a third.

Two-Person, Non-Zero-Sum Game Experiments

One reason for studying experimental games is that they are interesting. When you spend a lot of time thinking about how people should behave in theory, you become curious about how they actually behave in practice. A second reason for studying experimental games is that you may gain insights that will enable you to play better. That is a much more important consideration in non-zero-sum games than it is in zero-sum games. In the two-person, zero-sum game, players can obtain the value of the game by their own efforts alone; they don't have to worry about what their opponent does. In the non-zero-sum game, unless you are willing to be satisfied with a minimal return—as a buyer and a seller may have to be when they fail to reach an agreement—you *must* be concerned with how your partner plays. Similarly, in a sequence of prisoner's dilemmas, what you anticipate will be your partner's play will affect your own.

Granted that it is worthwhile to learn more about how people behave, why should we turn to the laboratory rather than to actual life for our data? Examples of non-zero-sum games in everyday life are certainly common enough. Lawrence E. Fouraker and Sidney Siegel, in *Bargaining and Group Decision Making* (1960), answer the question thus:

> In the specific case of Bilateral Monopoly, it would be extremely unlikely that appropriate naturalistic data could be collected to test the theoretical models. This is not because the phenomenon is unusually rare. Indeed, there are numerous daily exchanges that are conducted under conditions that approximate Bilateral Monopoly: a franchised dealer negotiates with a manufacturer regarding quotas and wholesale price; two public utilities bargain about the division of some price they have placed on a joint service; a chain grocery store negotiates with a canner, who in turn must deal with farmers' cooperatives; labor leaders in a unionized industry deal with management in that industry; and so forth.

The trouble with "real" games is that they are not set up for our convenience. The variables are not controlled; we are not likely to find two situations that are identical except for one variable. This makes it difficult to determine just how important a variable is in influencing the final outcome. Also, it is usually not feasible to determine the payoffs. In the laboratory, on the other hand, the players can be separated to avoid personal contact (an unnecessary complicating factor), the payoffs are clear, and the variables can be altered at will. It is also possible to motivate the players by making the payoffs sufficiently large—in principle, at least.

Some Experiments on the Prisoner's Dilemma

The many experiments conducted on the prisoner's dilemma all share a common purpose: to determine under what conditions players cooperate. Among the significant variables that determine how a player will behave are the size of the payoffs, the way the other person plays, the ability to communicate, and the personality of the players. In a series of experiments by Alvin Scodel, J. Sayer Minas, David Marlowe, Harvey Rawson, Philburn Ratoosh, and Milton Lipetz that were described in three articles in the *Journal of Conflict Resolution* between 1959 and 1962, a prisoner's dilemma game and some variations on it were played repeatedly. Let us consider some of the observations of the experimenters.

The game shown in figure 5.21, which I will call game 1, was played fifty times by each of twenty-two pairs of players. *C* and *NC* are the cooperative and noncooperative strategies, respectively. The players were physically separated throughout the fifty trials, so no direct communication was possible. At every trial, each player knew what his or her partner had done on every previous trial.

At each trial, each player had two choices, making a total

Figure 5.21

GAME 1

	C	NC
C	(3, 3)	(0, 5)
NC	(5, 0)	(1, 1)

of four possible outcomes. If the players had picked their strategies at random, we would have expected the cooperative payoff (3, 3) 25 percent of the time; the uncooperative payoff (1, 1) 25 percent of the time; and one of the mixed payoffs (5, 0) or (0, 5) 50 percent of the time. In fact, the uncooperative outcome predominated. Of the twenty-two pairs, twenty had more uncooperative payoffs than they had any other combination. Even more surprising, the tendency of the players was to be more uncooperative as the game progressed.

Game 1a was a repetition of game 1, with one variation: the players were allowed to communicate in the last twenty-five of the fifty trials. As you would expect, the results on the first twenty-five trials were pretty much the same as before. On the last twenty-five trials, there was still a tendency not to cooperate, but it was not so pronounced as when the players couldn't communicate.

Game 2 had the same payoff matrix as games 1 and 1a, but a variation was introduced. The subjects didn't play against each other but against the experimenter, though they weren't aware of it. The experimenter played in accordance with a predetermined formula: at each trial, he did what the subject did. If a subject cooperated, the experimenter did also (on the same trial), and the subject received 3. If the subject didn't cooperate, he received only 1. The game was repeated fifty times, just as games 1 and 1a were. The players chose the uncooperative strategy 60 percent of the time and played noncooperatively in the second twenty-five trials more often than they did in the first twenty-five trials.

You would think that in fifty trials the subjects would realize that they were not playing against just another person. And, presumably, subjects who caught on would play strategically. Both postexperimental interviews and the experimental results indicated, however, that every subject believed the responses of his or her "partner" to be legitimate. Those players who noticed a similarity between their own and their "partner's" play attributed it to coincidence.

In several other prisoner's dilemma games, the same pattern was repeated. In game 3 (see figure 5.22), for example, a player only gained 2 by defecting from the cooperative outcome. Nevertheless, 50 percent of the time, in the first half of a thirty-trial sequence, the players were uncooperative. And in the last fifteen trials, the percentage went up to 65 percent. When the experiment was repeated with the second player replaced by the experimenter (who always played uncooperatively), the frequency of cooperative play was virtually unchanged. In game 4 (see figure 5.23), a prisoner's dilemma in which the payoffs were mostly negative and the players had to scramble to cut their losses, there was virtually no cooperation at all.

Figure 5.22

GAME 3

	C	NC
C	(8, 8)	(1, 10)
NC	(10, 1)	(2, 2)

Figure 5.23

GAME 4

	C	NC
C	(−1, −1)	(−5, 0)
NC	(0, −5)	(−3, −3)

Some of the most interesting experiments were not really prisoner's dilemma games at all. Three of these games are shown in figure 5.24. Each of the three games was played in thirty-trial sequences, and in each game it was surprising how frequently the players failed to cooperate. In game 5, the players failed to cooperate 6.38 times on the first fifteen trials

Figure 5.24

GAME 5

	C	NC
C	(6, 6)	(4, 7)
NC	(7, 4)	(−3, −3)

GAME 6

	C	NC
C	(3, 3)	(1, 3)
NC	(3, 1)	(0, 0)

GAME 7

	C	NC
C	(4, 4)	(1, 3)
NC	(3, 1)	(0, 0)

(on average) and 7.62 times on the last fifteen trials: a small but statistically significant increase.

Games 6 and 7 turned out about the same. In game 6, the players were uncooperative slightly more than half the time and were a little more cooperative during the first half of the trials than they were in the second half. In game 7, the players cooperated about 53 percent of the time, but in the last fifteen trials *they failed to cooperate more than half the time.*

Throughout these experiments, there was a marked consistent tendency to play uncooperatively. Uncooperative play is understandable in games of the prisoner's dilemma type, where it has obvious advantages, at least in the short run. But the tendency persisted into the last three games, and this is much more difficult to explain.

In game 5, uncooperative play was rewarded only part of the time: a player gained only if the partner cooperated—and not very much at that. And if the partner played uncooperatively also, the player received the smallest possible payoff. In games 6 and 7, it was absurd to play uncooperatively. In game 6, there was no chance of gaining, and some chance of losing, if you played uncooperatively, and in game 7, a player who failed to cooperate always received a smaller payoff, whatever the partner did. The partner's play affected only the amount that the player lost. Despite this, in every game but the last, uncooperative play predominated. Even in the last game, the outcome was very close to what you would expect if the players had picked their strategies by tossing a coin. Moreover, as the game progressed, the tendency to cooperate became weaker rather than stronger.

Why players fail to cooperate is not altogether clear. A player might want to exploit a partner or might fear that a partner is about to exploit him or her. A player might not understand what the game is all about or might doubt that the partner does—though this last possibility doesn't seem very likely. If a player with insight into the game fails to cooperate because she fears her partner won't "read her message," she would presumably cooperate when she was permitted to talk to her partner and make her message plain. But, in fact, there was only a slight increase in cooperation when the players were allowed to communicate.

Not only were the players slow to cooperate with each other; they appeared to be almost oblivious to what their partners did. They were not even suspicious when their partner's play was identical to their own. Suspicious or not, their playing uncooperatively 60 percent of the time for fifty-trial periods in the face of this behavior borders on the incredible. *There was roughly the same amount of uncooperative play when the experimenter duplicated the subject's play as there was when the experimenter always played uncooperatively.*

The subjects seemed to regard these games as purely competitive: beating one's partner was most important, and the player's own payoff was only secondary. This tendency to compete, which has been observed by many experimenters and which increased as the game progressed, has been attributed to boredom and the smallness of the monetary payoffs. One wonders if this determination to beat one's partner would diminish if an appreciable amount of money were at stake.

Brian Forst and Judith Lucianovic (1977) also reported on a prisoner's dilemma "game," one played with real prisoners. They compared the fraction of codefendants who either pleaded, or were ultimately found, guilty with that of solo defendants and found the two were substantially the same. But despite the realistic trappings, they doubted that the situation really was a prisoner's dilemma since the accused usually managed to communicate even if both were in jail while awaiting, and during the trial, and there was a very real prospect of later retaliation by the betrayed partner.

A Bidding Experiment

In 1963 James Griesmer and Martin Shubik ran some experiments using as their subjects Princeton undergraduates. The basic experiment went like this:

Two players simultaneously picked a number between 1 and 10. The player who picked the higher number received nothing. The player who picked the lower number received that amount (in dollars or dimes, depending on the experiment) from the *experimenter*. If both players picked the same number, they tossed a coin and the winner of the toss took that number. If, for example, one player picked 5 and the other picked 7, the player who picked 5 received 5 units from the experimenter, and the other player received nothing. If both players picked 5, a coin was tossed and the winner of the toss received 5 while his partner received nothing. After each trial, the players were informed what their partners had done, and the game was repeated. The players were separated throughout the series and could not consult.

This game is almost identical to the gasoline price war example discussed earlier. By the same induction argument used before, we can "deduce" the superiority of the most uncooperative strategy: a bid of 1. If both players adopt this strategy, the payoffs will be very small—on average 1/2. If one player is competitive and the other is not, the competitive player will generally win something, but not very much. The greatest rewards clearly go to the players who cooperate. If both players keep their bids high, the joint profit will be greatest, and though one of the players will necessarily get nothing on any particular trial, repetition of the game gives both players a chance to do well.

If both players in a pair realize that they will do best if they cooperate, they still have the problem of figuring out how to synchronize their bidding, since no direct communication is possible. The most direct way to cooperate, and the way that maximizes the joint expected profit, is to bid 10 all the time. This yields an average return of 5 to each player. The one flaw in this plan is that a player can still get nothing, since in the case of ties the payoff is determined by chance. At a small cost in expected value, the players can raise their assured profit considerably. On the first of each pair of trials, each player picks a 10. The player who won the toss on the first trial picks another 10 on the second trial, while his partner chooses a 9. Thus, on every pair of trials, one player always wins 10 and the other always wins 9. The average return to each player is 4 3/4, and the guaranteed return is 4 1/2.

As it happens, almost all the players approached the game competitively; they tried to outsmart one another rather than jointly exploit the experimenter. Some players made cooperative overtures by bidding 9's and 10's alternately, but these were most often misinterpreted as attempts to lull the partner into a false sense of security. (This was later confirmed by the subjects in explicit statements and was obvious from their behavior during the play.) A few of the pairs did manage to get together, and when they did, it was invariably by alternating bids of 9 and 10, so that each won 9 on every other trial. This is not the most efficient way to cooperate, of course, but it is fairly good for subjects playing for the first time.

One of the things the experimenters wanted to study was the "end-effect"—uncooperative play on the last of a series of trials—but no evidence of it was observed during the experiment. When players managed to get together, they played cooperatively throughout the series of trials. If anything, there was more cooperation at the end than at the start. The experimenters tried to isolate end effect by telling some pairs the exact number of trials they would play and not telling others. Presumably the end effect would occur when the players knew which was the last trial, but in fact that there was no difference in play. Perhaps the payoffs were too small to motivate defection. Generally, two players who "found each other" were so pleased that they left well enough alone. (When the players were uncooperative from the first—and most were—the situation didn't arise.)

One variation of the game is worth describing in detail because it illustrates a point I made before in connection with utility theory. In one experiment, the basic game was played in sets of three trials, with one additional rule: if in the first two games a player received nothing, in the third game the player automatically received whatever he bid. This meant that a player could *always* get a payoff of 10 by bidding 10 on all three trials.

Players who were competitive in the earlier games usually played competitively in this game as well. Each of the players made low bids during the first two trials, and each won one of the first two games. The situation then was exactly the same as if they had been playing an ordinary game, since the new rule was not applicable. The players realized that if they played competitively on the last trial, as they had before, their profit, even if they won, would be considerably less than it would have been if they had played differently from the start. The realization that they could have done better apparently affected their utility functions, for they bid much more in the third game than they had earlier—in what essentially were identical situations.

An Almost Cooperative Game

There is another kind of cooperative, two-person game in which the basic problem is coordinating both players' strategies in accordance with their mutual interest. In games in which the players cannot communicate directly, Thomas Schelling (1958) suggested, the players must look for certain clues to help them anticipate what their partners will do. The clue could be a past outcome, for example, or it might be suggested by the symmetry of the payoff matrix. Richard Willis and Myron Joseph (1959) ran experiments to test Schelling's theory that prominence was a major determinant of bargaining behavior. In all, three different matrices were used. I call them 1, 2, and 3 (see figure 5.25).

The players were divided into two groups. Group *A* played game 1 repeatedly and then switched to game 2, which it also played repeatedly. Group *B* started with game 2 and ended with game 3. No communication between players was allowed during the play.

Figure 5.25

GAME 1

(10, 20)	(0, 0)
(0, 0)	(20, 10)

GAME 2

(10, 30)	(0, 0)	(0, 0)
(0, 0)	(20, 20)	(0, 0)
(0, 0)	(0, 0)	(30, 10)

GAME 3

(10, 40)	(0, 0)	(0, 0)	(0, 0)
(0, 0)	(20, 30)	(0, 0)	(0, 0)
(0, 0)	(0, 0)	(30, 20)	(0, 0)
(0, 0)	(0, 0)	(0, 0)	(40, 10)

Clearly, the players must solve the problem together. Unless they pick the same row and column, neither player will get anything. As a secondary goal, each player may try to reach the most favorable outcome of those that lie along the diagonal.

In game 1, Schelling offers us no clue as to what the proper play should be. In game 3, also, no clear-cut outcome suggests itself, but one of the two middle strategies would seem more likely than the extreme ones. It is only in game 2 that symmetry dictates a clear choice: the (20, 20) payoff corresponding to each player's second strategy.

What actually happened was quite surprising. When group *A* played game 1, there was a battle of wills in which each player fought for dominance. Each player played the strategy that would give him a payoff of 20 if the other yielded. After the initial battle, in which there were few agreements, one player finally gave way and that team settled down to an equilibrium point. When they switched to game 2, their play was strongly influenced by what had happened before. Rather than go to the (20, 20) payoff, which symmetry would suggest, these players went most often to one of the asymmetric equilibrium points. Three quarters of the time, the player who was dominant in game 1 remained dominant in game 2 too.

The actions of group *B* were even more unexpected. They started with game 2 and arrived at an agreeable equilibrium point much faster than group *A* had, which is not too surprising. But the agreement tended to be at one of the extremes—the first row and column or the third row and column—rather than on the middle row and column as suggested by symmetry and Schelling. As before, when the players in group *B* switched to game 3, the dominance that had been established in the earlier game prevailed.

Generally, repetitive play—picking the same strategy again and again—was the signal that was used to suggest an outcome. The repetitive play continued when it struck a responsive chord in the partner and even, often, when it didn't. The most equitable outcome in games 1 and 3 would be a synchronized, alternating scheme in which one partner is favored in one play and the other is favored in the next. This never happened, however, probably because such an arrangement is too intricate to arrange without direct communication.

Mother Nature as a Strategist

We observed earlier that game-theoretical models have a remarkable versatility. A model fashioned for one purpose often turns out to serve several others as well. A surprising example of multipurpose modeling is the application of game-theoretical concepts to evolution and ecology.

Game theory is usually thought of as a tool used by a thinking person involved in a game with other thinking people. But in a fascinating article, John Maynard Smith (1978) described some very unusual applications of game theory in which organisms "choose" very sophisticated strategies that enable them to survive as a species. These strategies are not chosen deliberately by individual organisms but are formed collectively by the entire species. It is as though there is an "invisible hand"—similar to the one that is understood to work in economics—that weaves individual behavior into a pattern for the entire species. These individual behavior patterns and their interactions may be described by payoff matrices; by analyzing these payoff matrices you can determine whether, and in what form, the species will survive.

The *fitness* of an individual organism is its ability to survive and have offspring—the ultimate goal in the survival game. The fitness of a species, also, is its ability to survive. These kinds of fitness are quite different and what favors one may not favor the other.

A potential conflict between these two kinds of fitness raises a basic question about the evolutionary mechanism. If the individuals of a species possess an altruistic trait, this may make the species fitter but it may make the individual that possesses it less so. A bird giving a warning cry when a predator appears helps the species to survive but, by calling attention to itself, hastens its own doom. If the bird remained mute, it would live longer and produce more offspring.

So the paradox comes to this: it is important for the survival of the species that individual members take certain risks, but it is the individual, not the species, that mates and passes on the traits it possesses. The bird that gave the warning cry will be dead, and its altruistic genes will be lost, while the selfish bird that stood mute will pass on its more prudent genes. Future generations of unwarned birds will pay the price for this temporary victory of opportunism, or so it appears at first.

But altruistic genes do survive and make the species fitter; and William D. Hamilton (1964) found the reason why. He observed that the critical factor in the survival of a gene isn't really the welfare of the organism that possesses it; rather it is the preservation of the gene itself and its replications. If a bird, taking a stroll with its family, loses its life because it gives a warning cry but saves ten of its offspring that would otherwise be killed, the loss of its own genes must be balanced against those of its offspring, but properly discounted.

The "discounting" process goes something like this: Since an offspring of the bird is as likely to inherit one parent's genes as the other, it will contain a particular gene of one of its parents only half the time. From the gene's point of view, the survival of its possessor is equivalent to the survival of two offspring. A sibling would also be worth half a gene, and more distant relatives would have even greater discounts. Hamilton added up the bits and pieces of the genes that were likely to be saved; if they outweighed the genes that were lost, he concluded the trait would survive.

In the game-theoretical survival model, then, the payoffs are not in terms of the individual's survival but in terms of the survival of the genes. A natural question to ask now is "Precisely which genes will survive?" And this is the question that John Maynard Smith attempts to answer.

Imagine a world in which all the members of a species have a particular trait. Subsequently there is a mutation—a small number of organisms are introduced with an alternative trait. What determines whether this new trait modification will die out or prosper? Smith has constructed a mathematical model that allows you to determine whether a trait is stable—that is, whether alternative traits will die out if they are introduced. He describes his model in the context of male animals competing for the attentions of a female.

There are many situations in which animals, like humans, have conflicting interests. When two animals compete for a mate or for territory, they generally begin by making threatening noises and taking an aggressive stance rather than by engaging in direct physical contact. If the confrontation appears to escalate into a physical struggle, they have a choice: they can back off, forgoing whatever spoils are involved but surviving for another day, or they can engage in serious physical combat. Smith calls animals that avoid combat *doves* and those that engage in combat, *hawks*. When a hawk engages a dove, it will gain whatever prize is at issue without even having to fight. But if a hawk encounters another hawk, one or the other is likely to be killed or seriously injured.

Now suppose that in a population composed solely of doves a mutation occurs and the population acquires a small number of hawks. Initially the mutants will thrive, because in confrontations between hawks and doves the doves will retreat and the hawks will mate; doves will only mate when there is no competition or when they are competing with another dove less persistent than themselves. But the early success of the hawks contains the seeds of their own downfall; as the hawks begin to proliferate, confrontation between hawks will become more common. To a dove, confrontation with a hawk is discouraging since the dove always leaves empty-handed; but to another hawk, such a confrontation can be a disaster resulting in death or serious injury. You might guess that at a certain point the doves and hawks will reach an equilibrium. This turns out to be the case. Smith assigned quantitative payoff values to the various possible confrontations, and on the basis of these payoffs, he tried to predict the long-run outcome.

Smith assumed that the animal that eventually mated received a payoff of +10 and an animal that was seriously injured received a penalty of −20. Two doves in combat do not hurt one another but since they do waste considerable time trying to bluff one another, Smith penalizes winner and loser alike by assigning them a payoff of −3. (Apparently when hawks engage either doves or other hawks, the matter is settled quickly.)

Putting these pieces together, Smith derives the payoff depicted in figure 5.26, which reflects the various possible confrontations.

Figure 5.26

	HAWK	DOVE
HAWK	(−5, −5)	(+10, 0)
DOVE	(0, +10)	(+2, +2)

When two hawks meet, one will gain 10 (+10) and the other will lose 20 (−20), so on average a hawk gets (−20 + 10)/2 = −5. When two doves meet, the successful dove gets +10 − 3 = 7 and the unsuccessful dove gets −3, so on average the payoff is 2. At a mixed confrontation the hawk gets 10 and the dove loses nothing.

In general when an organism of type X meets an organism of type Y the *expected payoff to* X is denoted by $E(X, Y)$; this payoff is the overall effect of wasted time, the risk of injury, and the chance of mating successfully and includes any other factor that is relevant to the spread of X's genes.

A gene I (or strategy I) is called *evolutionarily stable* if, whenever an alternative mutant, J, is introduced, I prevails and J dies out.

According to Smith's formal criterion, strategy I is evolutionarily stable if, for any alternative J, either condition 1 or condition 2 is true:

Condition 1: $E(I, I)$ is greater than $E(J, I)$
Condition 2: $E(I, I) = E(J, I)$ and $E(I, J) > E(J,J)$

Smith's reasoning goes like this. If a mutant, J, is introduced into a population in which I predominates, both I and J types will primarily be confronting I's. If an I does better against an I than a J does against an I, the I population will grow faster than the J population. This is expressed by condition 1. If both types do equally well against type I, both types will grow at the same rate and eventually there will be an appreciable number of type J organisms. If I's do better against J's than J's do against each other, this advantage will be critical and again, the growth rate of the I's will be greater than that of the J's. This is the effect of condition 2.

Let's see how this is applied to the conflict of hawks and doves. Suppose that a mutant dove is introduced into a population consisting only of hawks. Once doves gain a toehold, they will grow faster than the hawks because $E(D, H) = 0$ is greater than $E(H, H) = -5$. But this doesn't mean doves are fitter than hawks. If a mutant hawk is introduced into a population consisting only of doves, the hawks would grow faster than the doves since $E(D, D) = 2$ less than $E(H, D) = 10$. Neither a population of all hawks nor a population of doves is evolutionarily stable.

Smith observed that if the population adopts a "mixed strategy" so that eight-thirteenths of its population is hawks while the remaining five-thirteenths of its population is doves, it will remain evolutionarily stable against invasions from either hawks or doves. If we call such a population M, then it turns out that $E(H, M) = E(D, M) = E(M, M) = 10/13$; that is, everyone does the same against the mixed population initially. But if hawks proliferate, condition 2 takes effect, and since $E(M, H) = -40/13$ is greater than $E(H, H) = -5$, the mixed population will prevail. Similarly, if the doves become too numerous, their numbers will be thinned since $E(M, D) = 90/13$ is greater than $E(D, D) = 2$. Instead of formulating this in terms of three different populations, you can also think of the hawks and doves as being in equilibrium; if any imbalance occurs, the original state will be restored.

It is not hard to verify that the mixed strategy just described is evolutionarily stable, but it is not clear how a population would implement it in practice. There are at least two possibilities, and either will serve. One possibility is that there is a hawkish gene possessed by eight-thirteenths of the population that prescribes that its possessors invariably act hawkish. The other five-thirteenths of the population would have a dovish gene, with an analogous effect on its possessors. On the other hand, it is conceivable that there be a single gene that every member of the population contains which prescribes hawkish behavior eight-thirteenths of the time and dovish behavior five-thirteenths of the time. Either population would be evolutionarily stable.

This analysis of evolutionary stable strategies was used to explain the behavior of male dung flies seeking a mate. It is well known—both to human biologists and male dung flies— that females lay their eggs in cowpats; male dung flies, hoping to mate, wait for them there. Since females prefer freshly made cowpats, the obvious strategy for the male is not to linger at any one cowpat but to stay only briefly and move on as new cowpats are formed. But if every male adopted the obvious strategy, competition would be intense: a male would do better to adopt the subtler strategy of remaining after the others have left and keeping the few female stragglers that happened along to itself. The fact is that *any* single pure strategy adopted by every male would be inefficient for the species as a whole and would lead to intense competition for some females and to the neglect of others. A much more efficient strategy—and one that has been adopted in practice (see figure 5.27)—is to stagger the departure times so that the bulk of the males leave early in search of the preponderance of females but others leave more gradually and mate with the few tardy females. Although this mixed strategy has been observed in practice, it is still not clear how the strategy is implemented—whether each fly has its own fixed strategy and there is a mixture of different types of strategies or whether all the flies do the same thing: choose their own mixed strategy.

The ability of organisms to vary their behavior is often

Figure 5.27

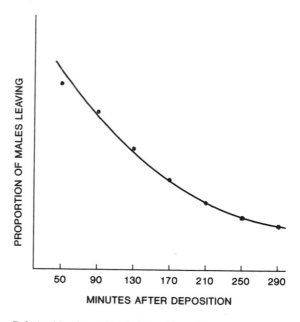

Relative Numbers of Male Dung Flies, *Scatophaga Stercoraria*, Leaving a Cowpat at Different Times After Its Deposition

NOTE: Reprinted, by permission of the publisher, from John Maynard Smith and G. A. Parker, "The Logic of Assymetric Contests," *Animal Behavior* 24 (1976):171.

beneficial to the species—this was true for male dung flies and it is also true in the hawk-dove model. One potential mutant in the hawk-dove model is particularly interesting; actually found in nature, it illustrates how nature resolves a paradox that plagues human beings, the prisoner's dilemma.

In the hawk-dove model, neither the hawkish nor the dovish strategy is completely satisfactory. The dove does well enough against other doves but does not get a fair share against hawks. Against doves, the hawk does as well as possible but confrontations with other hawks lead to disaster. What is needed is some sort of mediation that avoids physical combat yet prevents a conciliatory attitude from being exploited. One way of safeguarding what you have and still avoid excessive physical combat is to adopt what is termed the bourgeois strategy.

Suppose that in a population of animals, each has a certain region that he considers his own territory. From time to time there are conflicts (about a mate, perhaps) and once again we have the responses discussed earlier: after the preliminary posturing, the animal can flee (the doves) or can fight (the hawks). To these two old strategies we consider a new one: the bourgeois strategy. Bourgeois animals act like hawks when they are on their own territory but like doves when they are not. Since territory belongs to only one animal, in a bourgeois-bourgeois conflict there will never be any escalation; the stranger always backs off. Bourgeois-bourgeois duels are settled rapidly, with the intruder getting nothing and the inhabitant getting $+10$; a payoff of $+5$, on average. In the payoff shown in figure 5.28, it is assumed that each bourgeois has an even chance of being in his or her own territory.

Since $E(B,B) = 5$ is greater than both $E(H, B) = 2.5$ and $E(D,B) = 1$, the bourgeois strategy, once established, is safe against both hawkish and dovish incursions. Also, $E(B, H) = -2.5$ is greater than $E(H, H) = -5$, and $E(B, D) = 6$ is

Figure 5.28

	HAWK	DOVE	BOURGEOIS
HAWK	(−5, −5)	(10, 0)	(2½, −2½)
DOVE	(0, 10)	(2, 2)	(1, 6)
BOURGEOIS	(−2½, 2½)	(6, 1)	(5, 5)

greater than $E(D, D) = 2$, so either a hawkish or a dovish population would be vulnerable to a bourgeois invasion.

The predominance of the bourgeois strategy over hawks and doves seems to be confirmed in practice. Smith (1978) describes two examples—one observed by Hans Kummer and the other by N. B. Davies.

In a species of baboon in which a male forms permanent ties with one or more females, Kummer put male A together with a female while male B looked on. When the three were later reunited, A's association with the female was not challenged by B. To avoid the possibility that A simply dominated B, several weeks later B was isolated with a female while A looked on. This time it was A who became reconciled to B's association with the female. Possession was the cue for settlement of disputes.

Davies noticed that the speckled wood butterfly also adopts the bourgeois strategy in territorial disputes. Male butterflies seek sunlit spots on the forest floor because this is where the females are generally found. There are never enough spots for all the males so they are constantly flying in the forest canopy, scouting for vacancies. When a stranger invades an occupied spot, both stranger and resident circle briefly upward. One continues upward and the other—invariably the original resident—settles back on the spot. Davies noticed that possession is not merely a matter of strength; most of the butterflies he observed eventually settled in somewhere. Again, it was possession, not strength, that resolved these confrontations.

If a butterfly was unchallenged in a sunspot for more than a few seconds, it then considered the spot its own. Davies then asked, What if each of two males considered itself the owner? To answer the question, Davies insinuated an alien butterfly into a spot that already had an owner and waited. When the two butterflies confronted each other, they joined in a spiral flight that lasted ten times longer than the earlier ones in which it was clear to both parties who belonged and who didn't. A butterfly was clearly willing to escalate its quarrel on what it believed to be its own turf.

Not all territorial disputes are resolved as neatly as this; brute strength may play a role. When male fiddler crabs were observed disputing a burrow, the possessor was generally victorious (in 349 out of 403 cases, Smith reports). But in the minority of cases where the wandering male was successful, he was stronger.

A Computer Simulation of an Evolution Model

A few years ago Robert Axelrod, a political scientist, devised two interesting experiments involving people and computers (1980 *a,b*). At first glance the experiments seem to have little to do with Smith's model, but, in fact, the experiments and the model are very closely related. It turns out that the experiments yield insights into a number of different aspects of the evolutionary process.

In each of these experiments a number of people who were familiar with game theory—among them evolutionary scientists, social scientists, mathematicians, and so forth, many of whom had published papers in the field—were invited to enter a prisoner's dilemma tournament. The particular prisoner's dilemma involved had the payoffs shown in figure 5.29.

Figure 5.29

		OPPONENT	
		COOPERATE	DEFECT
PLAYER	COOPERATE	(3, 3)	(0, 5)
	DEFECT	(5, 0)	(1, 1)

The participants played a round-robin tournament in which every entrant played every other entrant a series of about 200 prisoner's dilemma games with the payoffs shown in the figure. The players also played one round with a player who played randomly and another with the mirror-image of themselves.

Once the tournament started, the players no longer participated directly. The players had to convey their playing strategies to a computer in the form of a program that was created in advance and that would remain unchanged throughout the tournament. The program indicated what the player wished to be done at each turn. This decision might be made on the basis of how the player's partner had acted earlier, it might be to act randomly, or it might invariably be to defect or cooperate. The players knew in advance that they would confront a random player, but they had no idea what the other player would do. (They did know that the other players were sophisticated, however.)

The players had a wide range of possible strategies from which to choose. At one extreme, they might always defect. While this strategy may seem intuitively unappealing, the arguments that support it are hard to refute even when the game is played repeatedly, as it was here. A defection strategy does best against strategies that ignore past experience—other purely competitive strategies, purely cooperative strategies, and random strategies. At the other extreme is the pure cooperative strategy, which always turns the other cheek. This strategy does fine when it encounters others of its kind but it doesn't do so well otherwise. All the other strategies fall between these two extremes.

Most of the strategies that were selected responded in kind to a cooperating partner—if not at first, eventually. The most important difference between the strategies was how they dealt with defecting partners. Some vindictive strategies defected and continued to defect until the end of the match if they observed a single defection by partner, whatever partner did after that defection. Some defected on the next play and then cooperated if their partner saw the light. And some continued to cooperate for a few plays, giving their partners a chance to reform before retaliating.

Some of the programs played a more dangerous game. They defected once or twice in the hope that they had a tolerant partner who would "educate" them before retaliating seriously; when they did this successfully, they picked up a few extra points. Some of the more cynical programs defected at the end of a run (in the first tournament, which was of fixed length), exploiting the fact that there are no rewards or punishments once the run is over. And finally, some strategies culled their partners' past history for cues to their future behavior. If their partners tolerated defections in the past, they would take a chance but not otherwise.

Considering the sophistication of the players and the variety of strategies available, the outcome of the experiments were surprisingly clearcut: "nice strategies"—those that never defect unless their partner defects first—invariably outperformed the others. There were no exceptions to this rule; each nice strategy outperformed every strategy that wasn't nice.

The nice strategies that did best punished defections but had short memories. They always gave their opponent a chance to reform. Strategies that irrevocably punished defectors for the rest of the run did not do nearly as well. When an unforgiving (but nice) strategy ran into an occasional defector trying to pick up a few extra points, there was a disaster for both. When the defector's misdemeanor was treated as a felony he lost heavily; but reform by the defector was precluded by the unyielding strategy of the partner so that ultimately the nice strategy lost as well.

The winner of both tournaments was a surprisingly simple strategy submitted by Anatol Rapoport called Tit-for-Tat. The strategy cooperated on the first play; thereafter it did on each play what its partner had done on the previous play. Tit-for-Tat had a number of weaknesses: for one thing, it never "noticed" that it was playing a random player so it kept up a futile attempt to reform it. (This strategy did worse against the random player than any other program.) But despite this blind spot, and despite the existence of competitive programs that had the same general philosophy but sophisticated variations as well, the Tit-for-Tat strategy was the clear winner in both tournaments.

The three properties that made it a successful strategy were niceness, forgiveness, and provocability: it never was the first to defect, it didn't hold a grudge too long, but it did not ignore its partner's defections. Tit-for-Tat is similar to the bourgeois strategy in Smith's model: it cooperated (the bourgeois deferred to its partner when it was appropriate), but it did not let defections pass unpunished (the bourgeois escalated when its partner broke the territorial rules). Axelrod felt that choosing a survival strategy in Smith's model was much the same as choosing a strategy in the computer tournament. In fact, the computer tournament might be considered a model for ecological development. ("Ecological" rather than "evolutionary" because no mutations are allowed in the computer tournament—all the strategies are there from the start.)

Imagine a population of animals in which pairs confront one another from time to time. Assume that each animal has a choice of two strategies—cooperate or defect—and that the payoffs are in terms of "expected number of surviving offspring" units. Imagine further a population in which each strategy is equally represented initially, and assume that the number of animals using each strategy in the next generation reflects the success that strategy had in this one (that is, the number of points it accumulated). Axelrod and Hamilton (1981) calculated what would happen if the strategies that were competing in the tournament met again and again, round after round. As the populations changed from generation to generation, a strategy's performance might change as well—what is fit in one population might not be in another. But in the main, it turns out, unfit strategies in the first round died out early and an equilibrium of sorts was achieved. By the five hundredth generation, eleven strategy groups had grown larger—the eleven strategies that had done best on the first round. Tit-for-Tat did best on the first round and continued to do best thereafter.

Assuming there is an analogy between surviving in a computer tournament and finding an effective evolutionary strategy, Tit-for-Tat's success in one environment suggests that it might be successful in the other as well. And it turns out that in the real world, it is, at least under some conditions.

Biologists have observed that many potential symbiotic relationships—relationships in which animals cooperate to their mutual advantage—exist in nature. But before these relationships can materialize, the same kinds of barriers to the cooperative solution discussed earlier, when players were people, must be overcome. Robert Axelrod and William D. Hamilton, an evolutionary biologist, discuss some of the barriers in their paper "The Evolution of Cooperation" (1981). Their statement—"The problem is that while an individual can benefit from mutual cooperation, each one can also do even better by exploiting the cooperative efforts of others. . . .

149

With two individuals destined never to meet again, the only strategy that can be called a solution to the game is to defect always despite the seemingly paradoxical outcome that both do worse than they could have had they cooperated" (p. 1391)—is a familiar story by now. When animals are unlikely to meet more than once, there is little motivation to cooperate, and Tit-for-Tat (or any other kind of cooperative strategy) is unlikely to thrive. When the prisoner's dilemma game is played repeatedly because the probability of meeting again is high, the Tit-for-Tat strategy comes into its own. And under these conditions, the strategy is often found in nature.

The variety of situations in which this kind of game-theoretical analysis can be applied is very surprising. For one thing, the organisms involved need not have a brain. As Axelrod and Hamilton point out, even bacteria have the capacity to play, because they can detect the response of a partner and then respond to their chemical environment. This is sufficient to allow them to punish defections and reward cooperation, because they have the ability to make their partners less fit, just as their partners have the same power over them. Moreover, their offspring can inherit this ability to discriminate.

The two organisms involved in a cooperative game don't even have to be able to "recognize" their partners; it is enough that they have continuous contact. Axelrod and Hamilton give a number of examples of such symbiotic relationships such as a hermit crab and its sea-anemone partner, a cicada and the microorganismic colonies of its bodies, and a tree and the fungi that inhabit it.

A more commonplace example of the Tit-for-Tat strategy in nature is the game played by the female fig wasp and a tree. The fig wasp sees its purpose as laying eggs; from the tree's point of view, the wasp's purpose is to pollinate flowers. If the fig wasp concentrates on laying eggs in a fig and neglects its pollinating duties, the tree cuts off the developing fig and all of its offspring die.

Axelrod and Hamilton cite another example of the cooperative game played by a small fish and its potential predator. The small fish eats parasites from the body, and even the mouth, of the large ones, while the large fish seeks its supper elsewhere. It is essential for such a relationship that there be consistent contact; this is managed by having an arranged meeting place. Such relationships have only been observed near the coast or on reefs but never in the open sea.

To maintain a cooperative relationship in a sequence of prisoner's dilemmas, there must be a good chance that defections will be punished and cooperation rewarded. In ant colonies that are stationary, symbiotic relationships are common; in honeybee colonies that are constantly moving, they are unknown. If there is a good prospect that the relationship will break up tomorrow, defection will become more likely today. Axelrod and Hamilton observe that seemingly harmless bacteria inhabiting the gut may suddenly become harmful when the gut becomes perforated. Other bacteria that are normally benign suddenly become dangerous when their host becomes sick or elderly. In each case the present reward outweighs the future prospects.

A final example is a game constructed by Martin Shubik as a casual amusement; it turned out to have several serious applications. Some time ago Shubik (1971) described the following parlor game: Auction off a dollar bill to an audience of several people but vary the usual rules. Allow the person who bids the most to pay what he or she bid and keep the dollar but require the person who made the *second highest bid* to pay what he or she bid without any compensation. Also, isolate the members of the bidding audience so that there is no collusion.

Shubik predicted that inexperienced players might bid

more for the dollar than it was worth. If your bid of 95¢ was raised to a dollar by a competitor, your choice would be to pay 95¢ without compensation (if the bidding dies at a dollar) or to bid $1.05 with the hope that your loss will be held to 5¢—the $1.05 you bid less the dollar you get. (The trouble is that your competitor may now bid on for the same reason.)

After Shubik made his predictions, Richard Tropper (1972) formally tested them. He had subjects participate in three different auctions (following Shubik's rules). In both the first and second auctions the "winning" bid exceeded the value of the prize. (In one case it was three times the size of the prize.) The subjects seemed to learn from experience, however, because in the last trial the bids were reduced substantially.

At first glance this game seems frivolous and without any application (not even to real auctions where only the winning bidder has to pay). In fact, however, it has several applications. When Shubik originally described this game, he mentioned that it was similar in many respects to an arms race. When two bidders in a Shubik auction have another round of bidding, both increase their potential losses but neither has increased his or her chance of winning the auction. In an arms race, similarly, the relative positions of both parties after they have increased their arms level may be the same yet they have both made a substantial extra investment.

The same model emerges in an evolutionary competition. John Haigh and Michael Rose (1980) use just such a model to describe a game involving two competing animals. Two adversaries interested in the same territory, or mate, resolve their quarrel by trying to outlast one another. They continue posturing until one loses patience; the more durable of the two obtains the reward but *both* pay the price in lost time.

After making certain quantitative assumptions about the payoffs, Haigh and Rose attempt to derive an evolutionarily stable strategy. It is pretty clear that no single waiting time is optimal, for if there were a population that consisted only of animals with a fixed waiting time, a mutant that waited just a bit longer would prove superior. The authors conclude (what nature concluded before them) that a mixed strategy is called for. An even better solution would be to pick out some assymetric feature of the situation—superior size, longest in the territory, superior strength, and so forth—and use that to decide the issue; that would save everyone's time.

Some General Observations

The purpose of studying experimental games is to isolate the factors that determine how players behave. Ultimately, we hope, enough knowledge will be obtained to allow us to predict behavior. So far, however, there has been only a limited amount of experimentation, and the results have not been entirely consistent. Much of the work has centered on the two-person, non-zero-sum game, focusing on the prisoner's dilemma and in particular on identifying the elements that determine whether a person will compete or cooperate.

Participants in a prisoner's dilemma game, it has been established, often fail to cooperate. Several possible reasons for this were mentioned earlier. Players may be interested only in furthering their own interests. They may be trying to do better than their partner (rather than maximizing their own payoff). They may feel that their partner will misunderstand cooperative overtures, or, even if the partner understands, may only exploit them. Or they may simply not understand the implications of what they are doing.

There is little doubt that players often fail to see all there is to see in a game. Rapoport (1962) has suggested that people fail to play minimax in zero-sum games (and they frequently do) because of lack of insight. There is some evidence that people fail to cooperate in non-zero-sum games for the same reason. Often players just do not conceive of playing other than competitively. In one set of prisoner's dilemma experiments, it was established by postexperimental interviews that, of twenty-nine subjects who understood the basic structure of the game, only two chose to play noncooperatively. In the bidding experiments conducted by Greismer and Shubik (1963 a,b,c) discussed earlier, many of the players were competitive because they weren't aware of all the possibilities. They perceived the game as being competitive and never thought about cooperating. And those who did cooperate felt uneasy—as if they were cheating the experimenter by conspiring illegally.

Despite all this, and even granting the truth of Rapoport's observations on zero-sum games, something more than the ignorance of the players seems to be involved when players fail to cooperate in non-zero-sum games. For one thing, highly sophisticated people often play uncooperatively, especially in one-trial games. For another, if the rewards are sufficiently enticing, either cooperative or competitive behavior can be induced. Apparently the factors that determine whether the play will be cooperative or competitive are quite complex and the experiments do not indicate the ratio of distrust, ambition, ignorance, and competitiveness that leads to competitive play.

Not all the significant variables in a game are controlled by the experimenter. One of the most important elements of the game is the personality of the players. It is to be expected that two people will react differently even in identical situations. If we are to have any chance of predicting the outcome of a game, therefore, we must look beyond the formal rules and into the attitudes of the players—a difficult job.

Experimenters have long been aware of the importance of personality and have tried to measure it or control its effect in a number of ways. The following clever scheme, for example, was devised by Jeremy Stone (1958) to measure the aspirations (or greediness) of players and their attitude toward risk.

Players were given a large number of cards and were told that they would be playing a series of games against an unnamed opponent. On each card were listed the rules of a particular game. A typical card might read: "You and your partner will each pick a number. If your number plus twice your partner's does not exceed 20, you will each get the number of dollars that you picked; otherwise you will each get nothing." Among the cards there would also be a game exactly like this one, but with the players' roles reversed. It would read: "You and your partner will each pick a number. If twice your number plus your partner's does not exceed 20, you will each get the number of dollars that you picked; otherwise you each get nothing." The cards were then paired, so that subjects played the role of both players—in effect, playing against themselves. The final score of a player was dependent on his or her own attitudes only.

Some experimenters tried, but failed, to correlate what players did in the prisoner's dilemma with their scores on certain psychological tests. Some connection *was* found between the players' political attitudes and how they played. In one study by Daniel R. Lutzker (1960), players were scored on an internationalism scale on the basis of how they responded to statements such as: "We should have a world government with the power to make laws which would be binding on all member nations," or "The United States should not trade with any Communist country." Lutzker observed a tendency of extreme isolationists to cooperate less in the prisoner's dilemma than did extreme internationalists.

Morton Deutsch (1960*b*), in a similar study, found that two characteristics related to the prisoner's dilemma were dependent on each other, and each of these, in turn, was dependent on the person's score on a particular psychological test. The traits Deutsch was concerned with are "trustingness" and "trustworthiness." To study these, a variation of the prisoner's dilemma was tried in which the play was broken down into two stages. In the first stage, one player picked a strategy. In the second stage, the other player picked a strategy, *after being informed what the first player had done*. The subjects played both roles, and in many different games. Players were *trusting* to the extent that they played cooperatively when they played first; they were *trustworthy* to the extent that they played cooperatively at the second turn in response to a partner's cooperative play. It turned out that, to a significant degree, "trustingness" and "trustworthiness" are positively correlated, and each trait is inversely related to authoritarianism (as measured by the F scale of the Minnesota Multiphasic Personality Inventory).

There have been attempts to relate other attributes of individuals, such as intelligence or sex, to the way they play, but with little or no success. The indications are that people's previous experience—and, in particular, their profession—affects their play. I have already mentioned how differently businessmen and students behaved in the last trial of a series of wholesaler-retailer games. Incidentally, a study was made of the prisoner's dilemma using prisoners as subjects. The results seem to contradict the adage about honor among thieves. Indeed, prisoners behaved very much like college students when playing this game: by and large, they tended not to cooperate.

Experimenters have also tried to control personality in games by imposing an attitude on the players. The players were told: "Do the best for yourself and don't worry about your partner," or: "Beat your partner." In this way, it was hoped, the players' attitudes would be fixed. These instructions had little effect, however.

Behavior Patterns

Certain consistent patterns have been observed by several experimenters who have studied sequences of prisoner's dilemma games. There is a tendency for players to become less cooperative rather than more cooperative as the games are repeated—though it is not clear why. Also, the reaction of subjects to their partner's behavior is in a way consistent. Suppose a sequence of prisoner's dilemmas is divided into two halves. Suppose, also, that in one case a player's partner always acts cooperatively in the first half, and in another case the partner always acts uncooperatively in the first half. (In such experiments, the "partner" is either the experimenter or an assistant.) Uncooperative playing in the first half of the games, it was found, is more likely than cooperative playing to elicit a cooperative response in the second half.

The effect of allowing the players to communicate is also subtler than might at first appear. Some experimenters found that the chance of a cooperative outcome was increased if the players were allowed to communicate. An experiment by Deutsch (1960a) indicated, however, that this was so only for individualistic players: players who wanted to win as much as possible and who didn't care how their partners did. The cooperative player who was also concerned about how a partner did, and the competitive player whose main concern was doing better than a partner, played the same way, whether they were able to communicate or not.

In another experiment carried out by Deutsch (1960a) to

analyze further the effects of communication, three basic situations were set up: a bilateral threat, in which each player had a threat against the other; a unilateral threat, in which only one player had a threat; and a no threat. (People have a threat if they can lower their partner's payoff without changing their own.) In each of these three situations, some players were given the opportunity to communicate and others were not. Players who had the opportunity to communicate chose not to do so, however. In one situation—the bilateral threat—there was more cooperation when the players could communicate but didn't than when they did. When the players did try to communicate, negotiations often degenerated into repetition of threats. In the unilateral case, on the other hand, if the players communicated they generally became less competitive; and when the players could communicate but didn't, conflicts between them were created or aggravated. In short, the effect of allowing communication depends on the attitudes of the players, and, in turn, the attitudes of the players may be affected by the ability to communicate.

The most significant drawback to these experiments is, again, the insignificance of the payoffs. This was manifested explicitly and implicitly in many ways. It was reflected in the increased competition: beating one's partner became more important than maximizing one's payoff. It was indicated by the players' own statements; they admitted playing frivolously during long, monotonous, prisoner's dilemma runs. It was shown by the general tendency (at least on the part of students) to be cooperative in the wholesaler-retailer game once a tacit understanding had been reached and by the speed with which defectors were punished. (It is easier to punish a greedy partner when it costs you $10 than when it costs $10,000.) My point is not that the experimenters erred, but rather that the results must be analyzed cautiously.

The Two-Person, Non-Zero-Sum Game in Practice

It is difficult to study "games" that are actually "played" in real life. Still, it has been tried. Determining how fruitful these attempts have been is better left to the experts. I will be content to say just a few words in passing.

One of the fields in which game theory has been studied is business, and the analysis has been essentially descriptive. One study that I referred to earlier concerned the tactics used in a taxicab rate war. Elsewhere, competition in business was compared with armed conflict, and the conditions that tend to precipitate price wars were pinpointed. In the field of advertising, more formal models have been constructed, encompassing many problems: setting the advertising budget, allocating funds to the various media and/or geographic areas, determining the best time slots to advertise in, and so forth. There has even been an attempt to fit Indian burlap trade into the framework of game theory, allowing for possible coalitions, strategies, and threats.

Political scientists have also borrowed concepts from two-person, non-zero-sum game theory. I mentioned Maschler's application to disarmament models. Other models have centered about nuclear deterrence and bomb testing. Some models have been studied in extensive form, with the "plays" being military resources: transforming nonmilitary resources into military resources, standing pat, and so forth. The role played by communication, especially between the United States and Soviet Russia, has also been studied on this basis.

In these models, it is virtually impossible to get an accurate quantitative expression that will reflect the payoffs. Still, enough insight may be obtained by assigning approximate figures to justify the studies. In the next chapter, on games in which there are more than two players, even more difficult quantitative estimates will be made—for example, the power of a member of a voting body will be examined.

Solutions to Problems

1. This game is the celebrated prisoner's dilemma, of course, and both the single-play and multiple-play cases were discussed in the text at length. There is really no clear answer to (a), (b), or (c), but Tit-for-Tat was the most successful strategy in the computer game.

2. a. You certainly can be better off if your strategies are restricted; a liquor retailer who lowers prices to meet the competition might well profit from a law that fixed prices.

Figure 5.30

		YOUR PARTNER	
		A	B
YOU	C	(1, 3)	(5, 1)
	D	(0, −90)	(3, −100)

b. If there is communication and the opportunity to form binding agreements, you can threaten to play D unless your partner agrees to play B; with no communication, your partner can play A and be reasonably certain you will play C.

c. In the symmetric game shown in figure 5.31, the outcome isn't clear. (It depends on whether you can communicate and give side payments and on your negotiating ability.) But if the rules of the game force you to move first and you play C, you will very likely get 30, your best possible payoff.

Figure 5.31

		YOUR PARTNER	
		A	B
YOU	C	(10, 10)	(30, 25)
	D	(25, 30)	(20, 20)

d.   It can be to your advantage or disadvantage to have your opponent know your utility function. If goods are sold by bargaining and you want something badly, you'd best hide your feelings; but, as mentioned earlier, a company having financial difficulties would want to inform a labor union with which it is bargaining of its troubles.

3.   The answer to all parts of this problem depend on your individual reactions; but there is no doubt that at some point, if your share is small enough, you will *not* try to maximize your profits but, instead, minimize your opponent's.

4.   As the matrix stands, it is cheaper for the community to ignore the laws then to enforce them completely. But if the community enforces the law 10 percent of the time, the payoff matrix can be depicted in figure 5.32. With enough publicity, this payoff will induce rational speeders (?) to slow down at a cost of only 2 to the community.

Figure 5.32

		COMMUNITY
		10% ENFORCEMENT
DRIVER	SPEED	(−10, −7)
	DO NOT SPEED	(0, −2)

5. a.   This is another form of the prisoner's dilemma. Notice that i dominates $i + 1$; that is, you do at least as well playing 9 as you do playing 10, at least as well playing 8 as you do playing 9, ... It would seem, therefore, that each player should play 1. But it's hard to accept an outcome in which each player gets 1/2 on average when each player could have received 5. And it's harder yet if the game is repeated fifty times.

b.   In an early experiment by James Griesmer and Martin Shubik (1963), players sometimes alternated 10's and 9's to signal their intent to cooperate. Their partners, more often than not, viewed this as a competitive game and felt their partners were either stupid or trying to deceive them; in a few cases the signal was read correctly to the benefit of both players.

GAME THEORY

6. I hope you haven't guessed it, but your partner was you. If you compare graphs *a, c; d, g; b, e;* and *f, h,* you will see that they are really the same game but from the point of view of different players. In effect, then, you were playing against yourself.

As a rough guide to how you did, I list the Nash value (obtained by maximizing XY) next to each of the paired games. If you received a lot of zeros, you're too greedy; if your scores were much lower (but positive) than the Nash sums, you're too modest.

GAME PAIR	SUM OF NASH VALUES
a–c	25 + 10
b–e	50 + 5
d–g	11.5 + 53.3
f–h	56.6 + 28.3
	239.7 for all 8 games.

This experiment was originally devised by Jeremy Stone (1958).

The n-Person Game

Introductory Questions

As we move from the earlier chapters to the later ones, players seem to lose control over their fate. In two-person, zero-sum games with equilibrium points, players could always exact their due and if there were no equilibrium points they could obtain it on average. In two-person, non-zero-sum games, they had to share control of their own fate with a partner but they had control over their partner's fate, which they could use as a threat. In n-person games, even this threat is generally denied the players. They must form coalitions with others and consider what inducements they must offer and accept. Once again, start by trying to answer these introductory problems.

1. An engineer, a lawyer, and a treasurer lose their jobs when their firm goes bankrupt. They are offered new jobs and an additional bonus if they join as a team. The additional bonus (in thousands of dollars) for the engineer-lawyer pair is 16, for the engineer-treasurer pair, 20, and for the lawyer-treasurer pair, 24. If all three join as a team, the bonus is 28.

Figure 6.1

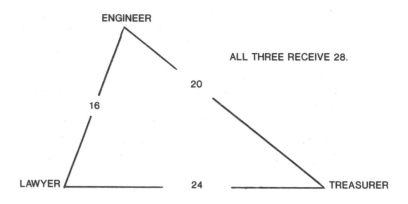

ENGINEER

ALL THREE RECEIVE 28.

20

16

LAWYER   24   TREASURER

Before joining, the team must decide how it will divide the bonus.

   a. Should/will all three join together (since they get the greatest bonus that way)? If they do join together, will every pair get at least as much as they would have alone? (If not, why shouldn't a pair break off from the third member and do better?)

   b. How should the players divide the bonus if each of the three pairs join?

   c. Answer questions (a) and (b) assuming that all three get a bonus of 40 instead of 28 but nothing else is changed.

2.   Three storekeepers, A, B, and C, with seasonal businesses share a single warehouse that only one uses at a time. Their storage needs vary. A needs $6,000 worth of space, B, $8,000 worth, and C, $10,000 worth. The warehouse need only be large enough to accommodate the largest user so it rents for $10,000. How should the renters allocate their costs to reflect their individual needs?

3.   In this chapter I will try to define a measure for "power." Using only your own intuitive notions of power, try to answer the following questions:

   a. A committee consisting of A, B, and a chairman, C, must pick one of three alternatives: X, Y, and Z. A prefers X

to Y and Y to Z; B prefers Z to X and X to Y; and C prefers Y to Z and Z to X.

    While the committee's decision is determined by majority vote, the chairman's vote is decisive if no single alternative is chosen by two members. Which member of the committee has the most power? If everyone is sophisticated, what should the outcome be? (Note: This question is subtler than it appears to be.)

b. In New York City's Board of Estimate, the mayor, controller, and council president each used to have three votes, the borough presidents of Brooklyn and Manhattan had two votes, and the other three borough presidents had 1 vote. Now all borough presidents have two votes and all other members have four votes. Bills are passed by majority vote. Which members have more power because of the change and which have less power?

c. In a certain voting body decisions are made by majority rule.

    (i) A small group forms a bloc; before each vote the group members retire to a back room, determine their own position by a majority vote, and then return to the parent body and vote unanimously on their position. Will this procedure change their power? How?

    (ii) Suppose that members have different numbers of votes (for example, they may be stockholders voting their number of shares), and a new member is added, increasing the total number of votes. Do the old members necessarily lose power? Can they actually gain power? (Suppose there were three members with thirteen, seven, and seven votes, respectively, and a new member is added with three votes.)

d. The number of votes that a state has in the electoral college reflects its representation in the House of Representatives and the Senate. Its membership in the House is roughly proportional to its population, but the two extra senators are a bonus given to large and small states alike. Do these extra two votes given across the board dilute the power of the larger states unfairly?

    e. In a body of five members in which decisions are made by majority vote, the members have one, two, three, four, and five votes, respectively. Comment on the relative power of these voters. Is their power proportional to the number of their votes?

4. A legislator who votes against her own inclination on one issue in return for support on one she considers more important is said to be logrolling. Logrolling disguises the preferences of voters and distorts the voting mechanism, but it must benefit those directly involved or they wouldn't do it. Is logrolling good or bad for society as a whole?

5. A voting body may make decisions in many ways—different members may have different numbers of votes, there may be runoff elections, and so forth. But certain basic rules hold for *any* reasonable procedure; would you include these among them?

    a. If a group prefers A to B (in a two-way vote) and B to C, the group should prefer A to C.

    b. If all members have a single vote, they should all vote for their most preferred alternative and the group should select the plurality preference.

    c. The group should never choose alternative A if there is an alternative B that would be preferred by the group in a two-way election between A and B.

6. The number of members in the House of Representatives from a state should reflect its population, but in general this would require that a state have a fractional member in the House—a practical impossibility. Consider the following approximation to the ideal: express the ideal representation of each state as a whole number plus a proper fraction. Give each state its whole number of representatives and in addition give the states with the highest fraction an additional member so that the total number of representatives from all states equals the size of the House. Can you see any reason to object to this plan?

7. Members of the town council are considering several proposed projects but have only the means to finance one of them. All agree that the council's decision should somehow reflect the feeling of its members and that everyone's pref-

erences should be weighted equally. What is your reaction
to each of the following plans?

a. Have the members cast a single vote for their pet project
and adopt the majority preference.
b. Have a "tournament" in which the projects are voted on
two at a time. At each stage eliminate the loser and adopt
the final survivor.
c. Have the members vote on their pet project and eliminate the least popular choice. Have a sequence of votes,
eliminating one option each time, and then select the
survivor.

The *New York Times Magazine* of October 20, 1968, carried an article entitled "The Ox-Cart Way We Pick a Space-
Age President," in which the role played by the electoral
college in presidential elections was examined. The author,
John A. Hamilton, asserted that voters in large states have an
advantage over voters in smaller states, despite the uniform
bonus of two senatorial votes which is allotted to large and
small states alike. In his words: "Under the winner-take-all
methods of allocating electoral votes the big states exercise
inordinate power. Although many more voters will go to the
polls in New York than in Alaska, they will have a chance to
influence many more electoral votes and, on balance, each
New Yorker will have a far better chance to influence the
election's outcome."

The advantage that the voter from a large state has is not
immediately obvious. The bloc of electoral votes controlled
by New York is clearly more powerful than that of Rhode
Island, but an individual voter in New York has less influence
on how the state will vote than the individual voter in Rhode
Island does, and it is the strength of the individual voter with
which we are concerned. Hamilton balances these two opposing tendencies and comes up with a plus for the large
states. For evidence, he looks to the record. He points out

that a switch of only 575 popular votes in New York in 1884 would have made James Blaine President instead of Grover Cleveland. He then lists four presidential elections in which 2,555 votes, 7,189 votes, and 10,517 votes in New York and 1,983 votes in California would have made the difference. On the basis of this he concludes that the large states have an influence on presidential politics that surpasses their actual size.

Granted that large states (or, rather, voters in large states) have the advantage in the electoral college, the question arises whether this is a peculiarity of the way the votes are distributed in our Congress or is always true in this kind of voting system. In general, how can we assess the "power" of an individual voter?

The fact is that voters in large states need not be more powerful than other voters. Suppose there are five states, four with a million voters and one with 10,000 voters. Now, whatever representation the state with 10,000 voters has in the electoral college (as long as it has *some* representation), it is as strong as any other state, since a majority must include three states—*any three states.* In effect, all states have equal power in the electoral college (assuming decisions are by majority vote and the large states all have the same representation). But an individual voter in a small state has much more influence on how his state will vote, and consequently is more powerful.

Suppose I modify the situation slightly. Assume that one state has 120 votes in the electoral college, three states have 100 votes, and one state has 10 votes. Now the 10-vote state is absolutely powerless, and so are the state's voters. No matter what the final vote is, the position of the 10-vote state is irrelevant, since it cannot affect the outcome by changing its vote; the voters in the small state are completely disfranchised.

The second problem—that is, actually assigning numbers to the players that will reflect their voting strengths—will be considered later.

The presidential election is itself a kind of game. The voters are the players, the candidates are the strategies, and the payoff is the election results. This is an n-person game, a game in which there are more than two players. The distinction is convenient, for games with three or more players generally differ radically in character from those with fewer than three players.

Of course, there may be considerable variation in n-person games also. Some examples follow.

Some Political Examples

In a local political election, the main issue is the size of the annual budget. It is well known that each voter prefers a certain budget and will vote for the party that takes a position closest to his or her own. The three parties, for their part, are completely opportunistic. They know the voters' wishes and want to gain as many votes as possible; they couldn't care less about the issues. If the parties have to announce their preferred budgets simultaneously, what should their strategy be? Would it make a difference if their decisions were not made simultaneously? Would it make much difference if there were ten parties instead of three? Two parties?

In 1939 three counties had to decide how some financial aid, allocated by the state, would be divided among them. This aid was to go toward school construction. In all, four different schools were to be built and distributed among the three counties.

Four different plans were proposed—I will call them *A, B,*

C, and D—each prescribing a different distribution. The final plan was to be selected by a majority vote of two of the three counties. The relationship of the counties, the plans, and the distribution of the schools are shown in figure 6.2.

Figure 6.2

PLAN

		A	B	C	D
	I	4	1	2	0
COUNTY	II	0	0	1	2
	III	0	3	1	2

The numbers in the figure indicate how many schools were to be built in each county according to each plan. Assuming that the plans were voted on two at a time until all but one was eliminated, which plan would win if each county backed the plan that gave it the most schools? Does it matter in what order the vote was taken? Would it have paid for a county to vote for one plan when it actually preferred another?

A nation that elects its governing body by a system of proportional representation has five political parties. Their strengths in the legislature are determined by the number of seats they hold: eight, seven, four, three, and one. A governing coalition is formed when enough parties join to ensure a majority; that is, when the coalition controls twelve of the twenty-three seats. Assuming that there are benefits that will accrue to the governing coalition—such as patronage, ministerial portfolios, and so forth—what coalition should form? What share of the spoils should each party get?

Some Economic Examples

In a certain city, all the houses lie along a single road. If it is assumed that customers always buy from the store nearest them, where should the retailers build their stores? Does the number of stores make any difference?

Several fashion designers announce their new dress styles on a given date, and, once announced, they can't be changed. The most important feature of the new dresses and the feature that determines which dress is finally bought is the height of the hemline. Each customer has her own preference and will buy the dress that best approximates it. Assuming the designers know the percentage of women who will buy each style, which styles should the companies manufacture?

An agent writes three actors that she has a job for two of them, any two of them. The three actors are not equally famous, so the employer is willing to pay more for some combinations than he is for others. Specifically, A and B can get $6,000; A and C can get $8,000; and B and C can get $10,000. The two that get the job can divide the sum any way they like, but before they can take the job, they must decide how to divide the money. The first two actors to reach an agreement get the job. Is it possible to predict which pair will get the job? How will they divide the profits?

A wholesaler wants to merge with any one of four retailers who jointly occupy a city block. If the merger goes through, the wholesaler and the retailer will make a combined profit of a million dollars. The retailers have an alternative: they can band together and sell to a realty company, making a joint profit of a million dollars that way. Can the outcome be predicted? If the wholesaler joins a retailer, how should they divide the million dollars?

171

An inventor and either of two competing manufacturers can make a million dollars using the patent of one and the facilities of the other. If the investor and one of the manufacturers should manage to get together, how should they share their profit?

An Analysis

One of the advantages of using a game as a model is that it permits you to analyze many apparently different problems at the same time. That is the case here. In the local political election, for instance, the parties are in a situation very similar to that of the retailers in the first economic example, and they in turn have the same problem as the fashion designers in the second economic example. In terms of the political example, the issue need not be quantitative, such as fixing a budget. It might involve taking a position on any problem where there is a more or less one-dimensional spectrum of opinion, such as liberal vs. conservative, or high tariff vs. free trade. At first it might seem that the parties should take a position somewhere in the middle; and if there are only two parties, and they make their decisions in sequence rather than simultaneously, this is generally what happens. In presidential elections, the conservative party and the liberal party often (though not invariably) tend to move to the center, on the theory that voters on the extremes have nowhere to go. Under similar circumstances, the competing stores will tend to cluster at the center of town; and the fashion designers will follow moderate tastes. When there are many political parties (or many stores in town, or many fashion designers), it may be better to cater to some outlying position. If there is proportional representation, one of ten parties may be willing to settle for 20 percent of the vote, even if it means giving up any chance of getting the votes at the center.

If you look at the problems that the players must solve in these games—how to win the most votes, how to get the most schools, how to sell the most dresses—you see that the primary concern is with power: the power of a player or a coalition of players to affect the final outcome of the game.

In the context of the n-person game, power, while real enough, is a subtler concept and more difficult to assess than in the simpler, one-person and two-person games. In the one-person game, players determine the outcome themselves or share control with a nonmalevolent nature. In the two-person, zero-sum game, a player's power—what the player is sure of getting on the basis of his or her own resources only—is a good measure of what the player can expect from the game.

The two-person, non-zero-sum game is a little more complicated. Here a player also wields the power to punish or reward a partner. Whether this power over the partner can be made to increase the player's own payoff, and to what extent, depends on the partner's personality. Since it cannot be so converted by the player him- or herself, its value is not completely clear. Nevertheless, this kind of power is highly significant and must be taken into account in any meaningful theory.

In the n-person game, the concept of power is even more elusive. Of course, there is always a minimum payoff that players can get by themselves. To get any more, players must join others, as they did in the two-person, non-zero-sum game. But in the n-person game, if other players fail to cooperate, the player has no recourse. It would seem, then, that beyond the minimum payoff the player is helpless. Yet coalitions of apparently impotent players do have power. Thus the player has potential power that requires the cooperation of others to be realized. Making this concept of potential power precise is one of the basic functions of n-person game theory.

To illustrate how the problem of evaluating a player's potential power might be approached, let us look at one of the earlier examples. Suppose that, in the third economic example, one of the actors approaches an outside party before the bargaining begins and offers him whatever earnings he receives in return for a lump sum. The third party would take over the bargaining for the actor: he could offer and accept any bargain he wished on the actor's behalf and, of course, would take the risk of being left out completely, with no gain at all. The amount that the third party would be willing to pay for the privilege of representing the actor might be considered an index to the actor's power in the game.

This states in a more specific way the problem of evaluating potential power but takes us no closer to a solution. Actually, many different solution concepts are possible for this kind of game, and they will be discussed later. For the moment, let us look at one approach. The sketch in figure 6.3 summarizes the game.

Figure 6.3

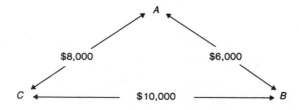

In this game, your first inclination might be to assume that B and C will join together, since they have the most to gain: $10,000. Just how they will divide this money is another matter. One suspects that player A, who is not even included in the coalition of *B* and *C*, will play an important part in determining how the money is divided, for if *B* and *C* should fail to reach an agreement, they would both have to look to *A*. The respective shares of *B* and *C* should be somehow related to the value of the coalitions *AB* and *AC*. And since the coalition *AB* has a smaller value than the coalition *AC*, it seems reasonable to conclude that *B* will get less than half of the $10,000.

This argument has an obvious weakness, however. Once *A* realizes that he has an inferior position, he is bound to lower his demands. It is clearly better for him to get into some coalition, even if he has to leave the lion's share to his partner, than to stand by himself and get nothing. Thus even the one "obvious" conclusion—that *B* and *C* will form the coalition—seems open to question.

This should give a rough, qualitative idea of how the problem might be attacked and that is enough for now. It might be interesting to anticipate a little, however, and look at some of the conclusions reached about this game on the basis of one theory, without dwelling on the reasoning behind the conclusions. I will use the Aumann-Maschler theory, because it is particularly simple.

In the three-actor example, the Aumann-Maschler theory does not predict which, if any, coalition will form. It does predict, however, what a player will get if he or she manages to enter a coalition; and this amount, at least for this game, does not depend on which coalition forms. Specifically, the theory predicts that *A* should get $2,000; *B*, $4,000; and *C*, $6,000.

In the wholesaler-retailer merger, the Aumann-Maschler theory does not predict what coalition will form (it never does). In fact, in this case it does not even predict the precise way the players will divide the million dollars. It simply states that if a retailer joins the wholesaler, then the retailer will get something between a quarter and a half million dollars. In the problem of the inventor and the two competing manufacturers, the theory predicts that the inventor will get virtually all the profit.

The school construction and the national election examples are two more illustrations of voting games—a common source of n-person games. The problem of assigning strengths to the parties in the legislature is very similar to that of assigning strengths to states in the electoral college (a problem discussed earlier). In the national election example, the real power of the players (that is, the parties) is not what it seems to be. In this instance, the party with four seats has exactly the same strength as the party with seven seats. The party with four seats, if it is to be a member of a majority coalition, must combine either with the eight-seat party (and possibly others), or with the seven-seat party and a smaller party (and possibly others). So must the seven-seat party (except that, in the second alternative, the roles of the seven-seat and four-seat parties are reversed).

Finally, in the school construction issue, the order in which the counties vote is critical. Suppose the order is *CABD*. This means that proposal A and C are voted on first; then B is pitted against the winner of that contest; and finally D is matched against the winner in the second contest. The outcome may be depicted as shown in figure 6.4. Thus, in an election between C and B, C would win; this may be indicated by $C > B$. Also, $D > A$, $C > A$, $D > C$, $B > D$, and $A = B$. (That is, there would be a tie between A and B.)

Figure 6.4

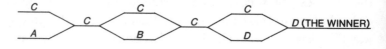

176

If the order is *DACB*, the outcome is as shown in figure 6.5. (Note that in the first order of voting, *CABD*, it is to county *I*'s advantage to vote strategically. In the second contest, county *I* should vote for *B* rather than for *C*, even though it prefers *C*. If it does vote for *B*, *B* will win the second vote and in the last contest *B* will win again. In this way, county *I* gets one school. If county *I* always voted blindly for the plan it preferred without regard for strategic considerations, plan *D* would be adopted and county *I* would have no schools at all.) Notice also that if *B* and *D* are matched first and *A* confronts the winner, *C* will be the final choice; if *D* meets *C* first and the winner meets *B*, *A* and *B* will tie. Since three of the four can win a clear victory and the fourth proposal, *A*, can be a cowinner with *B*, it is clear that the voting order is critical.

Figure 6.5

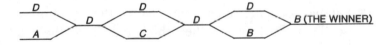

Admittedly, these illustrations are highly simplified. Reality is, almost invariably, much more complicated. One senator, for example, may be much more powerful than another, despite the fact that each has one vote. A phenomenon as complex as the election of the President of the United States is almost impossible to capture in a single model. Consider some of the difficulties.

In the process of electing a president, many games are played. One of them takes place at the party national convention. Among the delegates are a small number of party leaders who control a large percentage of the votes and who act (they hope) so as to maximize their political power. These party leaders must make two basic decisions: which candidate to support, and when to announce their support. The party leaders have some a priori preferences among the aspirants, but they must be cautious, because they lose everything if they back a losing candidate. If they manage to back the eventual winner, they reap certain benefits, depending on the strength of the leader, the candidate's need for their support, and so forth. These benefits may take the form of reciprocal support or patronage.

Two political scientists, Nelson W. Polsby and Aaron B. Wildavsky (1963), constructed a model of a convention that took into account such phenomena as: (1) the bandwagon flocking to an aspirant once his success becomes obvious; (2) the inverse relationship between an aspirant's chances and the concessions he must yield to get support; (3) an aspirant's need to stimulate the expectation of victory. If an aspirant doesn't win or increase his vote within a certain period of time, his vote thereafter will not remain constant; it will drop. As a result, aspirants often "hide" votes they control and release them slowly when they think they'll do the most good.

Once the candidates are nominated, there is a new game—very similar to the one described in the first political example. Now the two parties must take a position not on a single issue but on many. Of course, they are not quite as free in doing so as was assumed in the example, since parties do have ideological commitments.

The voter, too, plays a game. Even in the simplest instance, where there are only two candidates, there may be problems. If voters' positions are at one end of the political spectrum, their wishes may be virtually ignored if they vote mechanically for the candidate whose position is closest to their own. If they withhold their vote, the election of the other, less favored candidate becomes more likely. Balanced against this risk is the chance of increasing the voter's influence in the future. The political parties will take the extreme voters more seriously if, whenever they move too far toward the center, there are wholesale defections. (When there are more than two serious candidates, the situation is even more complicated, for then you must decide whether to vote for a very desirable candidate with a doubtful chance of winning or for a less desirable candidate with a somewhat better chance of winning.)

Obviously it is out of the question to analyze all these games in their original form. Instead, we must take simple models and try to formulate a solution that seems plausible. This may be done, and in fact has been done, in several ways. Some solutions aim at reaching an arbitration point based on the strengths of the players. Others try to find equilibrium points such as exist in a market of buyers and sellers. Some define a solution as sets of possible outcomes satisfying certain stability requirements.

The von Neumann–Morgenstern Theory

In their book, *The Theory of Games and Economic Behavior* (1953), John von Neumann and Oskar Morgenstern first defined the *n*-person game and introduced their concept of a solution. All the work that has been done on *n*-person games since has been strongly influenced by this now classic work. It will be easier to understand the von Neumann–Morgenstern approach (from now on I will refer to it as N-M) if I discuss it in the context of a specific example.

Suppose that three companies—*A, B,* and *C*—are each worth a dollar. Suppose that any two, or all three of them, can form a coalition. If such a coalition forms, it will obtain an additional $9, so that a two-person coalition will be worth $11: the dollar each company had originally, plus the additional $9. And the three-person coalition will be worth $12. We assume that each company is completely informed and, for the sake of simplicity, that utility is identical to money. What remains is to determine the coalition that will form and the way the money will be divided. Before attacking this problem, however, I will make some general observations about n-person games.

The Characteristic Function Form

The game that I just described is said to be in *characteristic function form.* With each coalition is associated a number: the value of that coalition. The value of a coalition is analogous to the value of a game in the two-person case. It is the minimum amount that the coalition can obtain if all its members join together and play as a team.

For many games, the characteristic function form is the most natural description. In a legislative body, for example, in which decisions are made by majority vote, the coalition values are obvious. A coalition that contains a majority of the players has all the power; a coalition without a majority has none. In other games—games in which the players are buyers and sellers in the open market, for instance—the value of a coalition may not be so clear. But N-M show that in principle this kind of game may also be reduced to characteristic function form, as follows.

N-M start with an n-person game in which each player picks one of several alternatives, and in consequence of these choices, there is some outcome: a payoff to each of the players. The strategies available to the players may be fixing a price or quantity, casting a vote, hiring a number of new salespeople, and so forth. Having thus set the stage, N-M ask what would happen if a coalition of players—let us call them S—decide to act in unison to get the biggest possible combined payoff they can get. What should the coalition S hope to obtain?

This problem, N-M observe, is really the same problem we faced in the two-person game. The members of S constitute one "player," and everyone else constitutes the other "player." As before, we may compute coalition S's maximum payoff, assuming that the players not in S act in a hostile manner. This figure, denoted by $V(S)$, is called the *value* of coalition S; the value of any coalition may be calculated in this way.

This procedure suggests the same question that was raised before: Will the players not in S really try to minimize the payoffs of the players in S? And N-M's answer is the same here as it is in the two-person game: they will if the game is purely competitive. For this reason, N-M assume that the n-person game is zero-sum; that is, if the value of any coalition S is added to the value of the coalition consisting of the players not in S, the sum will always be the same. (If more than two coalitions form, the sum of the coalition values may decrease, but it will never increase.)

Superadditivity

Since there are many different kinds of n-person games, the values assigned to the coalitions may take on almost any pattern—almost, but not quite. A basic relationship exists between the values of certain coalitions that is a consequence of the way these values are defined.

Suppose R and S are two coalitions that have no players in common. A new coalition is formed that is composed of all the players either in R or in S; the new coalition is denoted by $R \cup S$ (R union S). Clearly the value of the new coalition must be at least as great as the sum of the values of coalition R and coalition S. The members of R can play the strategy that guarantees them $V(R)$, and the players in S can play that strategy that guarantees them $V(S)$. Thus $R \cup S$ can get at least $V(R) + V(S)$. (It is quite possible, of course, that $R \cup S$ can do even better.) This requirement, which must be satisfied by the characteristic function, is called *superadditivity*. Stated another way, a characteristic function is superadditive if, for any two coalitions R and S that have no players in common, $V(R) + V(S) \leq V(R \cup S)$.

Returning now to the original example, consider some of the outcomes. One possibility is that all three players will unite. In that case, symmetry would suggest that each of the players receive a payoff of 4. I will denote such a payoff by (4, 4, 4), the numbers in parentheses representing the payoffs to companies A, B, and C, respectively. Another possibility is that only two players will combine—say B and C—and share their \$11 equally, giving the third player, A, nothing. In this case, the payoff would be (1, 5 1/2, 5 1/2), for A would still have \$1. A third possibility is that the players are not able to come to an agreement and remain as they were originally, the payoff being (1, 1, 1). To see whether any of these outcomes, or some other outcome, is likely to materialize, let us imagine how the negotiations might go.

Suppose someone starts by proposing a payoff of (4, 4, 4). This seems fair enough. But some enterprising player, say A, realizes that she can do better by joining another player, say B, and sharing the extra profit with him. The payoff would then be (5 1/2, 5 1/2, 1). This is a plausible change. Both A and B would get more than they do in the earlier (4, 4, 4) payoff. C will be unhappy, of course, but there isn't much he can do about it—at least, not directly. But *C* can make a counteroffer. He might single out *B* and offer him $6, take $5 for himself, and leave *A* with $1—for a payoff of (1, 6, 5). If *B* accepts *C*'s counteroffer, it is then *A*'s turn to fight for her place in the sun.

Of course, this hopping about from payoff to payoff can be endless, for every payoff is unstable in that, no matter what payoff is being considered, there are always two players who have the power and motivation to move on to another, better payoff. For every payoff there are always two players who together get no more than $8; these two players can combine and increase their joint profit to $11. Obviously, then, this approach doesn't work.

Imputations and Individual Rationality

When people are first introduced to *n*-person games, they are tempted to look for a best strategy (or a best set of equivalent strategies) for each player and a unique set of payoffs that they might expect clever players to obtain; in short, a theory much like that of the two-person, zero-sum game. But it soon becomes clear that this is much too ambitious. Even the simplest games are too complex to permit a single payoff. And if a theory were constructed that predicted such a payoff, it would not be plausible or a true reflection of reality, since there are usually a variety of possible outcomes whenever an actual game is played. This is so no matter how sophisticated the players. There are simply too many variables—the bargaining abilities of the players, the norms of society, and so forth—for the formal theory to accommodate.

One thing that can be done, however, is to pare down the number of possible payoffs by eliminating those that clearly wouldn't materialize. This is what N-M do first. The N-M theory assumes that the ultimate payoff will be *Pareto optimal*. (An imputation, or payoff, is Pareto optimal if there is no other payoff in which all the players simultaneously do better.) On the face of it, this seems reasonable. Why should the players accept an imputation of (1, 1, 1) when all three players do better with an imputation of (4, 4, 4)? N-M also assume that the final imputation will be *individually rational*. That is, each player will obtain, in the final imputation, at least as much as he or she could get by going it alone. In our example, this would mean that each player must get at least 1.

Coalitional Rationality and the Core

When deciding which imputations of a game are plausible, you might also be tempted to demand that the imputations be *coalitionally rational*. An imputation is coalitionally rational if the members of each coalition receive a total payoff at least as great as their value. Why would any coalition take less than its members could get acting alone?

The trouble is that there may be no coalitionally rational imputation. If (a, b, c) were coalitionally rational, then $a + b \geq 11$, $a + c \geq 11$, and $b + c \geq 11$; adding both sides of the inequality and dividing both sides by 2, we obtain $a + b + c \geq 16\ 1/2$, which is impossible since the value of the three-person coalition is only 12.

The coalitionally rational imputations (if there are any), taken together, constitute the *core*. If a game has no core, it is unstable in the sense that whatever the payoff, some coalition has the power and motivation to break up the imputation and go off on its own. If the game discussed earlier (pp. 179–80) were altered so that the three-person coalition had a value of 20 instead of 12, an imputation would be in the core if each player received at least 1, each pair received at least a total of 11 (and, of course, the total payoff to all three players were 20).

Domination

Let's go back to the original bargaining, but now we will assume that the only proposals to be considered are individually rational, Pareto optimal imputations. If a proposal is made—that is, a coalition along with an associated payoff—under what conditions will an alternative proposal be accepted in place of the original?

The first requirement is that there is a group of players strong enough to implement the alternative proposal—an agreement to bell the cat is worthless if there's no one around capable of doing it. In addition, the players who are to implement the new proposal must be properly motivated. This means that all players must get more than they would if they stuck with the old proposal. If both these conditions are satisfied, if there is a coalition of players with both the ability and the will to adopt the new proposal, we say that the new proposal *dominates* the old one and call the implementing coalition the *effective set*.

To see how this works in the example on pp. 179–80, suppose the original coalition consists of all three players, with a payoff of (5, 4, 3). The alternative payoff (3, 5, 4) is preferred by *B* and *C* since they each get an extra dollar. Since *B* and *C* can get as much as $11 acting together, they can enforce the new payoff. (They could, of course, limit *A*'s payoff to 1, but they don't have to.) So (3, 5, 4) dominates (5, 4, 3), with *B* and *C* as the effective set. On the other hand, if

(1, 8, 3) were the alternative payoff to (5, 4, 3), it would not be accepted. Any two of the three players have the power to enforce this payoff, but no two are so inclined. *A* and *C* both prefer the original payoff, and *B,* who wants to change, doesn't have the power to bring it off alone. An alternative payoff of (6, 6, 0) would be preferred to the original by both *A* and *B.* But in order to get a payoff of $12, all three players must agree, and *C* is not about to. *C* can get 1 by going it alone and is not likely to settle for less.

A convenient way to illustrate the imputations is based on an interesting geometric fact about equilateral triangles: for any two interior points, the sum of the distances to the three sides is always the same. In our example, a payoff is an imputation if everyone obtains at least 1 and the sum of the payoffs is 12. All possible imputations can be represented, then, by points in an equilateral triangle that are at least one unit from each side. In figure 6.6, the imputations are represented by the points in the shaded region, and the point *P* represents the imputation (2, 2, 8).

In figure 6.7, the point *Q* represents the payoff (3, 4, 5). In

Figure 6.6

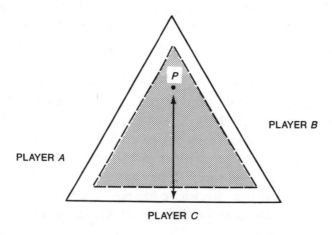

Figure 6.7

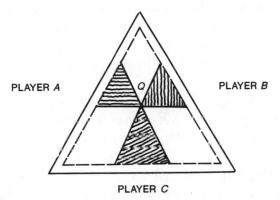

the horizontally shaded area are all the imputations that dominate Q with effective set BC. (The farther away an imputation is from the side marked "Player A," the greater the payoff to A in that imputation.) Both B and C get more in any imputation in the horizontally shaded area than they get at Q. The vertically shaded area and the diagonally shaded area represent those imputations that dominate Q with effective sets AC and AB, respectively. The plain areas contain the imputations that Q dominates, and the boundary lines represent the imputations that neither dominate nor are dominated by Q.

The von Neumann–Morgenstern Concept of a Solution

If you were asked to pick out a single imputation as the predicted outcome of a game, the most attractive candidate would seem to be that imputation which is not dominated by any other. There is a problem, however. There needn't be just one undominated imputation; there may be many. Worse still, as we have already observed, there may be no undominated imputation at all. Such is the case in this example. Every imputation is dominated by many others. In fact, the domination relation is what mathematicians call intransitive. Imputation P may dominate imputation Q, which in turn may dominate imputation R. And imputation R may dominate imputation P. (Of course, the effective sets must be different each time.) That is why the negotiations went round and round, without settling anywhere.

From the start, N-M gave up any hope of finding a single-payoff solution for all n-person games. There might be particular games in which such a solution would be plausible, but "the structure . . . under consideration would then be extremely simple: there would exist an absolute state of equilibrium in which the quantitative shares of every participant would be precisely determined. It will be seen, however, that such a solution, possessing all necessary properties, does not exist in general" (p. 34).

After ruling out the possibility of finding a single, satisfactory outcome for all n-person games, N-M assert that the only reasonable outcomes are imputations and go on to define their concept of a solution: "This consists of not setting up one rigid system of apportionment, i.e., imputation, but rather a variety of alternatives, which will probably all express some general principles but nevertheless differ among themselves in many particular respects. This system of imputations describes the 'established order of society' or 'acceptable standard of behavior' " (p. 41).

A solution, then, consists of not one but many imputations that together have a certain internal consistency. In particular, a solution S is some set of imputations that have two essential properties: (1) No imputation in the solution is dominated by any other imputation *in the solution*; (2) every im-

putation that is not in the solution is dominated by an imputation that is in the solution.

This definition of a solution "expresses the fact that the
standard of behavior is free from inner contradictions: no imputation y belonging to S [the solution]—i.e., conforming
with the 'accepted standard of behavior'—can be upset—i.e.,
dominated—by another imputation x of the same kind." On
the other hand,

the "standard of behavior" can be used to discredit any non-
conforming procedure: every imputation y not belonging to S
can be upset—i.e., dominated—by an imputation x belonging to
S. . . . Thus our solutions S correspond to such "standards of behavior" as have an inner stability: once they are generally accepted they overrule everything else and no part of them can be
overruled within the limits of accepted standards. (p. 41)

In general, there are many different solutions to any particular n-person game, and N-M do not try to single out a
"best" one. The existence of many solutions, they feel, far
from being a defect of the theory, is in fact an indication that
the theory has the flexibility necessary to deal with the wide
diversity one encounters in real life.

A related question—do solutions always exist?—is more
serious. For N-M, the question was crucial:

There can be, of course, no concessions as regards existence. If it
should turn out that our requirements concerning a solution S
are, in any special case, unfulfillable—this would necessitate a
fundamental change in the theory. Thus a general proof of the
existence of solutions S for all particular cases is most desirable.
It will appear from our subsequent investigations that this proof
has not yet been carried out in full generality but that in all
cases considered so far solutions were found. (p. 42)

Since this was written, there have been many attempts to
prove that solutions exist for all n-person games. They were

all fruitless until 1967, when William F. Lucas structured a ten-person game for which there was no solution, and the twenty-year-old question was finally settled.

N-M's concept of a solution can most easily be explained by means of an example. Suppose A, B, and C are players in a three-person game in which any coalition with either two or three players can get 2 units and a player alone gets nothing. This game has many solutions—an infinite number, in fact—but we will look at just two of them.

The first solution, which consists of only three imputations—$(1, 1, 0)$, $(1, 0, 1)$, and $(0, 1, 1)$—is indicated in the diagram in figure 6.8. In order to prove that these three imputations taken together are really a solution, two things must be verified: there must be no domination between imputations in the "solution," and every imputation outside the "solution" must be dominated by an imputation within it. The first part is easy enough. In passing from one imputation in the "solution" to another, one player always gains 1, one player always loses 1, and one player stays the same. Because a single player is not effective, there can be no domination. (In order to be in the effective set, players must gain in the change. It is not enough that they not lose.)

Also, in any imputation outside the "solution," there would be two players who get less than 1. This follows from the fact that no player gets a negative payoff, and the sum of the payoffs is 2. It follows that this payoff would be dominated by the payoff in the "solution" in which both these players received 1. The two players who originally received less than 1 would of course be the effective set. This shows that the "solution" is in fact a solution.

Another solution would consist of all imputations in which one player—say, player A—received $1/2$. I leave the verification of this to the reader. Figure 6.9 shows the solution.

The first solution may be interpreted in the following way.

Figure 6.8

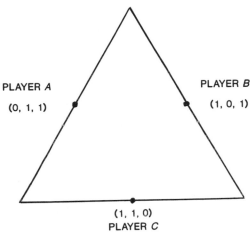

In every case, two players will get together, divide their 2 equally, and leave nothing for the third player. (Nothing is said about *which* two will join, however.) The payoffs are, of course, Pareto optimal. (They are also the most efficient, in that the gain per player is 1, while in a coalition of three the gain per player would be only 2/3.) They are also enforceable. This is called the *symmetric solution,* because all the players have identical roles.

In the second solution—a *discriminatory solution*—two players join, give the third player something less than his or her "fair share" of 2/3, and take the rest for themselves.

Figure 6.9

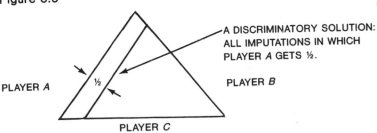

191

Which of the players join, and what is given to the third player, is determined by such factors as tradition, charity, fear of revolution, and so forth. Once the amount to be given to the "outsider" is determined, the game degenerates into essentially a two-person bargaining game, with the outcome depending on the personality of the players, and therefore indeterminate. All possible divisions between the two players, then, are included in the solution.

Some Final Comments on the N-M Theory

In order to construct a theoretical model of a real game, it is generally necessary to make some simplifying assumptions. The N-M theory is no exception. For one thing, N-M assume that the players can communicate freely; that is, they are able to communicate and/or act together as they please. Ideally all the players can communicate simultaneously; in practice, of course, they can't. And the deviation from the ideal that occurs in practice is very important. It has been shown experimentally that the physical arrangement of the players affects the bargaining, and players who are aggressive and quick to make an impact do better than others who are more reticent.

In addition, N-M assume that utilities are freely transferable between the players. If the payoff is in dollars, for instance, and A pays B one dollar, A's gain in utiles is the same as B's loss in utiles. This is a severe restriction on the theory and probably its weakest link. Actually, the assumption is not so restrictive as it may seem: it *does not* require that A's pain at losing a dollar be the same as B's pleasure in getting it. Also, there are many ways of assigning utilities to each of the players. However, if n people are in a game, it is unlikely that appropriate choices of utility functions could be made that would satisfy the restriction.

The Aumann-Maschler Theory of n-*Person Games*

The Aumann-Maschler theory (which I will denote A-M) is similar to the N-M theory in that it also uses the characteristic function form description of the *n*-person game, but in almost all other respects it is quite different. Suppose that *A*, *B*, and *C* are players in a three-person game in which the values of coalitions *AB*, *AC*, and *BC* are 60, 80, and 100, respectively; the value of the three-person coalition is 105; and the value of every one-person coalition is zero. The game is illustrated in figure 6.10.

Figure 6.10

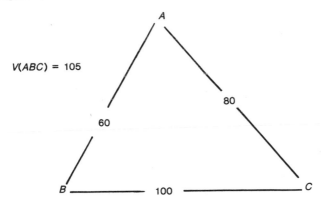

$V(ABC) = 105$

The A-M theory does not attempt to predict which coalition will form; its purpose is to determine what the payoffs would be once a coalition is formed. The theory takes into account only the strengths of the players; all considerations of fair play and equity are put aside. Before explaining precisely

193

what this means, let us assume that players A and B have tentatively agreed to form a coalition and are now concerned with dividing their payoff of 60. Their conversation might go something like this:

Player B (to player A): I would like my payoff to be 45, and so I offer you 15. I have a stronger position than you do in this game, and I think the payoffs should reflect it.

Player A: I refuse. I can certainly get 40 by going to player C and splitting the coalition value evenly with her. She would accept my offer, since at the moment she is getting nothing. But I'm not greedy. I'll accept an even split with you.

Player B: You're being unreasonable. No matter what I offer you, even if I offer you the whole 60, you can always threaten to go to player C and get more. But your threat is based on an illusion. At the moment C is in danger of getting nothing, so she's receptive to any offer that comes her way. But remember that defecting to C is a game I can play too. Once our tentative coalition disintegrates, you won't have C to yourself. You'll have to compete with me, and I have the advantage. If I join C our coalition would be worth 100, whereas yours would be worth only 80. If your objection to the 45–15 split were accepted as valid, then no agreement between us would be possible. If it ever comes to a fight between us for a partnership with C, we would each stand a chance of being left out entirely. I think we'd both be better off it you accept my reasonable proposal.

At this point, let us leave the bargaining table and take a closer look at the players' arguments. B's initial proposal seems unfair, since it prescribes that A get a much smaller share than B, and from one point of view it should be ruled out. But in the A-M theory the element of equity is not taken into account. If B's proposal is to be eliminated as a possible outcome, it must be on other grounds. (If a payoff is not precluded, that does not mean that it is *the* predicted outcome. A-M, like N-M, recognize that there may be many possible payoffs.)

A's objection—that he can go to *C* and offer her more than she would get from *B* (which is nothing) and still have more left over for himself than *B* offers him—is more pertinent. In effect, *A* is saying that if he and *C* have the power to form a new coalition in which they both get more than *B* would give them, there must be something wrong with *B*'s proposal. Specifically, *A* feels he isn't getting enough.

In his answer, *B* puts his finger on the flaw in *A*'s argument, which is that it goes too far. If *A*'s argument is allowed to rule out *B*'s proposal, it must rule out every other proposal as well. Even if *B* yielded the entire 60 to *A*, he would be open to the same objection, since *A* could still offer *C* 10 and receive 70 for himself. And, for that matter, so could *B*. If *any* coalition is to be formed, it must be in the face of this objection.

Granted, then, that *A*'s objection is not valid, and granted that *B* has a competitive advantage with *C*, why should *B* get 45? Why not 50? Or 40? A-M take the view that *B* should not get 45, and the reasoning goes something like this.

Suppose *A* persists in demanding more than 15 and *B* refuses to yield. Then *A* and *B* will have to compete for *C* as a partner. Presumably *B* would demand 45 from *C* also, for he wants 45 and it's all the same to him where he gets it. But if *B* gets 45 in the coalition *BC*, that leaves 55 for *C*. *A* can offer *C* 60 and take 20 for himself, which is 5 more than he was offered by *B*. Of course, *B* can always lower his demands, but then why ask so much from *A* to begin with? It turns out that the "proper" payoff when coalition *AB* forms is 40 for *B* and 20 for *A*.

As another illustration of this kind of argument, consider what happens when a three-person coalition is formed. A-M maintain that the payoffs to A, B, and C should be 15, 35, and 55, respectively. To see why, suppose we start with some other payoff—say 20, 35, 50—and see what goes wrong.

Since C is the one who seems to have been slighted, she should be the one to object, and so she does. She offers B 45, takes 55 for herself, and A has no recourse. There is only one way A can match C's offer: he must also offer B 45, and to do this, he has to lower his original demand of 20 to 15. A-M interpret this to mean that A was too ambitious in the first place. Of course, the payoff suggested by A-M, (15, 35, 55) is open to the same objection. A might, for instance, offer B 37 and take 23 for himself. But now, without raising her original demand of 55, C can outbid A by offering B 45.

This discussion gives a rough indication of the considerations that underlie the A-M theory. The next step is to reformulate in a more precise way what has just been said.

When the A-M theory of n-person games was first conceived, it was hoped that it would enable one to determine which coalitions would form and which payoffs would be appropriate. For tactical reasons the first problem was postponed, and A-M addressed themselves to the second: Given a particular coalition structure, what should the appropriate payoffs be? It was anticipated that once a payoff could be associated with each potential coalition formation, it would be possible to predict which coalition would form. So far, however, nothing has been done along this line. The fact is that you often can't predict which coalitions will form, even when you know (or think you know) how each coalition would distribute its payoffs. To see why, let's go back to our last example, in which the two-person coalitions had values of 60, 80, and 100. In this game, according to A-M, the payoff of a player is independent of *which* coalition he or she enters. The only thing that matters is that the player joins a coalition. That is, A receives 20, B receives 40, and C receives 60 if they enter *any* two-person coalition. (If they form a three-person coalition, they each get 5 less.) Consequently each person is indifferent as to whom he or she joins, as long as the player joins someone, and it is impossible to say, on the basis of the formal rules alone, that any two-person coalition is more likely to form than another.

The Formal Structure

A-M start by assuming the existence of a coalition structure in which each player is in precisely one coalition—possibly a one-person coalition consisting of only the player him- or herself. (They do not go into the question of whether the coalition is advisable or likely to form.) They assume (with no loss of generality) that all one-person coalitions have a zero value. Then a tentative payoff is assigned to each of the players, subject to certain constraints: the sum of the payoffs in any coalition should equal the value of that coalition; the coalition doesn't have the power to get any more, but it can see to it that it doesn't get any less. In addition, no player ever receives a negative payoff, for if he or she did, the deprived player would prefer to stand alone.

With these preliminaries out of the way, let's go back to the basic question: When is the payoff appropriate for the given coalition structure? In order to answer this question, A-M first look for the answer to another, subsidiary question: Is there any player with a "valid" *objection* against another player in his or her own coalition?

GAME THEORY

The Objection

An objection of player I to player J is simply a proposal that a new coalition be formed, call it S, with certain payoffs. Both players must be in the same coalition in the original structure. Otherwise I would have no claim over J. If the objection is to be valid, player I must be in S but player J must not; each player in S must get more than he or she was getting originally; and the sum of the payoffs of all the players in S must equal $V(S)$. The reason for all these conditions is clear: if the objection serves to show J that I can do better without him or her, I can hardly ask for J's cooperation in forming a new coalition. Also. the players in S must be paid more, or they'd have no motivation to join. Finally, $V(S)$ is all that S can be sure of getting, and there's no reason why S's players should settle for anything less. In effect, I's objection to J is the same that A made to B in the earlier example: "I can do better by joining S. The other prospective members of S also do better, so I'll have no trouble convincing them to join me. If I don't get a bigger share in the present coalition, I'll go elsewhere."

If, for a given payoff, no player has a valid objection to any other, A-M consider the payoff acceptable for that coalition structure. But what of the converse? If a player does have a valid objection to another, should it necessarily rule out the payoff? A-M think not. Otherwise in some games every pay-off in every coalition structure would be ruled out (except where only one-person coalitions are formed and objections are impossible). As a matter of fact, the original example was just such a game. So that there can be *some* acceptable pay-offs, A-M, under certain conditions, will allow the original payoffs to stand if J can truthfully counterobject: "I too can do better by leaving the coalition and joining some of the other players."

At first glance *J*'s reply does not seem to be really responsive to *I*'s objection. After all, if they both *can* do better, perhaps they both *should* do better. *J*'s answer seems to confirm what *I* suggested: the original structure should be broken up. There are two difficulties, however. In the first place, the "others" that *I* and *J* are threatening to join may be the same player(s). (In the example, she was.) And so *I* and *J* can't do better simultaneously. Also, when the original coalition dissolves, *I* and *J* are forced to be dependent on other players, so they have not necessarily improved their position. A-M assume that the members of a coalition would prefer to keep control over their own fate and will look elsewhere for partners only if some player persists in making demands that are disproportionate to his or her power. If a player can do better by defecting and a partner cannot, A-M feel that the second player is asking too much. But if *both* players can do better by defecting, it may behoove them both to sit still rather than go round and round and risk being left out of the final coalition. A-M consider a payoff to be acceptable, or *stable*, if, whenever a player *I* has a valid objection to *J*, *J* has a *counterobjection* to *I*.

The Counterobjection

A *counterobjection*, like an objection, is a proposal for a coalition *T*, with a corresponding payoff. It is made by a player *J* in response to an objection directed toward him or her by another player *I*. *T* must contain *J*, but not *I*. The point of the counterobjection is to convince the objector that he or she is not the only one who can do better by defecting, and it is essential that *J* make it plausible that such a coalition can form. To induce the players in *T* to join him or her, *J* must offer them at least what *I* offered them, if they happen to be in the objecting set *S,* and if not, what they would have obtained originally. But the sum of the payoffs must not exceed what *J* can afford to pay, *V(T),* the value of the new coalition. Finally, *J* must receive at least as much as he or she would have obtained originally.

The set of stable payoffs for a given coalition structure—which I will call the A-M solution—is called the A-M *bargaining set.*

A few points in the A-M theory should be stressed. For one thing, stable payoffs need not be fair. A-M are not seeking equitable outcomes but outcomes that are in some sense enforceable. Suppose, for instance, that the two-person coalitions have values of 60, 80, and 100, just as before, but the three-person coalition has a value of 1,000 rather than 105. Suppose further that a three-person coalition forms with a payoff of (700, 200, 100). Judging from the values of the two-person coalitions, *C* seems to be stronger than *B,* and *B* seems to be stronger than *A.* Nevertheless, A-M consider this payoff—in which the "weakest" player gets the most and the "strongest" player the least—stable. The reason for this is that *B* and *C,* despite their complaints, have no recourse. They can break up the coalition, of course, but then they would lose too, along with *A.* It is true that *A* would lose more, but the A-M theory does not recognize dog-in-the-manger tactics. For players' complaints to be considered valid, they must be able to do *better* elsewhere. Thus A-M avoid making interpersonal comparisons of utility, such as comparing a loss of 700 by *A* with a loss of 100 by *C.*

The Aumann–Maschler and von Neumann– Morgenstern Theories: A Comparison

The most important difference between the A-M and N-M theories is in their concept of a "solution." For N-M, the basic unit is a set of imputations. An imputation, taken alone, is neither acceptable nor unacceptable; it can be judged only in conjunction with other imputations. In the A-M theory, an outcome stands or falls on its own merits.

In the last sentence I referred to an outcome rather than an imputation, and this is another difference between the two theories: A-M do not assume the outcome will be Pareto optimal. On the face of it, it may seem absurd for players to decide on one outcome when with another outcome they all do better. Absurd or not, however, this is what often happens. Moreover, non-Pareto-optimal outcomes are important in the A-M theory. A-M are very much concerned with determining which coalitions and payoffs will hold under the pressure of players trying to improve their payoffs. The only weapon that players have is a threat of defection, and if this is to be plausible, defections must occasionally occur. Often the new coalitions are not Pareto optimal. Workers must strike occasionally, then, if threats are to be taken seriously, and when they do, the outcome is generally not Pareto optimal.

Perhaps the biggest advantage of the A-M theory is that it does not require interpersonal comparisons of utility. The payoffs may be given in dollars without affecting the theory; all that is required is that a person prefer a larger to a smaller amount of money. The A-M theory also drops the assumption of superadditivity, but this is less important. It is still a reasonable assumption but is no longer needed.

The A-M solution, like the N-M solution, sometimes consists of more than one payoff (for a given coalition structure), and the theory makes no attempt to discriminate between them. In the example in which the three-person coalition received 1,000, there are many stable payoffs for the coalition. Threats play a very limited role—they preclude only the most extreme payoffs—and the game is one of almost pure bargaining.

For a given game, there are generally many outcomes that are consistent with both the A-M and the N-M theories. But, in a sense, a much greater variety of N-M solutions is possible, and this makes the N-M theory much more comprehensive. In the three-person game, for example, in which any coalition with more than one player received 2 and single players received nothing, each coalition structure had only one stable payoff. The N-M symmetric solution for this same game consists of the three stable A-M imputations, and there is nothing in the A-M theory that corresponds to the N-M discriminatory solution. The greater variety of solutions in the N-M theory leads to greater flexibility but makes the theory almost impossible to test in practice. The A-M theory at times seems to narrow but is much easier to test. Consider the following example:

An employer wishes to hire one or two potential employees, A and B. The employer together with the employee will make a combined profit of $100. Together the employees get nothing, and the employer by herself gets nothing. All three may unite, and then the profit will also be $100.

In the A-M theory there are, basically, two possible outcomes: either the employer forms a coalition (with one or both employees) and gets $100, or she doesn't and gets nothing. The workers get nothing in any case. If one worker offered to take $10 and give the employer $90, the employer could counter by offering the other worker only $5 and the first worker would have no recourse. Thus any payoff in which a worker receives anything is unstable.

A-M solutions are very convincing in certain situations.

When there are a large number of workers who cannot communicate easily, they may very well compete in the way suggested by the A-M theory. But at other times it doesn't seem to apply. It is fairly obvious that unbridled competition hurts all the workers, and, not surprisingly, in real life workers often join together to act as a single coalition even though they gain nothing immediately but bargaining power. In effect, the game is reduced to a two-person bargaining game, with the workers acting as a single player. (In large industries, such as steel or motors, companies generally do not use wage levels to compete for the best worker; nor do workers offer to take a cut in salary to get a job. Both industry and labor merge into what Galbraith calls countervailing forces, and the result is in effect a two-person game.)

The N-M theory, on the other hand, has many solutions for this game. One consists of all imputations in which each worker gets the same amount. This may be interpreted as follows: the two workers join and agree to split anything they get equally, and they negotiate with the employer for $100. This results in a two-person bargaining game in which anything can happen. It is interesting that the sole payoff dictated by the A-M theory—$100 for the employer and nothing for the workers (assuming *some* coalition forms)—is contained in every N-M solution.

The core plays a special role in both the A-M and N-M theories. Since imputations in the core are undominated, the entire core is in *every* N-M solution, and every imputation in the core must be stable in the A-M sense since no objection is ever possible.

One last illustration of A-M's concern with enforceable rather than equitable outcomes follows.

Each of two retailers *A* and *B* has a customer for a certain item who is willing to pay $20, and each of two wholesalers *C* and *D* has a source at which the item can be obtained for

$10. In the four-person game consisting of A, B, C, and D, the value of a four-person coalition is $20; the value of any three-person coalition is $10; the value of the two-person coalitions AC, AD, BC, BD is each $10. The value of every other coalition is zero.

Now consider all the coalition structures in which the players get $20, the greatest amount possible. There are three possibilities: either of two two-person coalitions: (AD and BC) or (AC and BD); or the four-person coalition. In each of the three instances, both wholesalers will get the same payoff and both retailers will get the same payoff; but, in general, the wholesalers will get different payoffs from the retailers. This seems odd when you consider that the wholesalers and retailers have completely symmetric roles. But A-M's reasoning is this: If A is getting less than B, A can join the wholesaler who is getting the lowest payoff and they will both do better. If both retailers and both wholesalers are getting the same amount, there is no way to form a new coalition in which all the players do better. (For the sake of simplicity, I have taken some liberties with the A-M theory as it was originally formulated. I omitted certain assumptions about coalitional rationality, for example, and restricted the size of the objecting and counterobjecting sets to a single player. This introduces some changes—for one thing, it assures at least one stable payoff for every coalition structure—but the spirit remains the same.)

The Shapley Value

Lloyd Shapley (1953) approached the n-person game in still another way. He looked at the game from the point of view of the players and tried to answer the question: Given the characteristic function of a game, of what value is that game to a particular player?

From what we have seen, predicting the outcome of an arbitrary *n*-person game on the basis of the characteristic function alone would seem to be a hazardous business. The personality of the players, their physical arrangement, social custom, communication facilities, and so forth all have some effect on the final payoff. Nevertheless, Shapley found a method for calculating the worth of a game to each player—generally called the Shapley value—on the basis of the characteristic function alone. This number was arrived at a priori, with all other relevant factors abstracted away.

Shapley's scheme is only one of many that could serve the purpose. Why use one and not another? Shapley justifies his choice as follows: he lists three requirements that he feels any reasonable scheme should satisfy, and then he goes on to show that his scheme satisfies these axioms, indeed, that his scheme is the *only* one that does. The critical requirements are these:

1. *The value of a game to a player depends only on the characteristic function.* This means the values are assigned without regard to the identities or traits of the players. In a bargaining game, for example, in which the two players get nothing when they stand alone but share something when they get together, their values would be the same.
2. *A payoff in which each player receives his or her value is an imputation.* Shapley assumes that rational players will form an imputation. (He also accepts N-M's assumptions of superadditivity and the transferability of utilities.) Since the sum of the payoffs is necessarily equal to the value of the *n*-person coalition (by the definition of imputation) and the value of the game to players is their average payoff (in some sense), it follows that the sum of all the values should also equal the value of the *n*-person coalition.
3. *The value to a player of a composite game is equal to the*

sum of the values of the component games. Suppose a group of players is simultaneously engaged in two games. Define a new game with these same players in which the value of a coalition is equal to the sum of the values that it had in the original two games. In this new game, each player has a Shapley value, and axiom 3 states that it should equal the sum of the players' Shapley values in the original two games. Figure 6.11 may make this clearer. Note that game 1 is a composite of games 2 and 3. Coalition *BC*, which has a value of 3 in game 2 and a value of 4 in game 3, for example, must have a value of 7 in game 1. Axiom 3 states that in such a situation the value of game 1 to a player must be the sum of the values of game 2 and game 3.

Figure 6.11

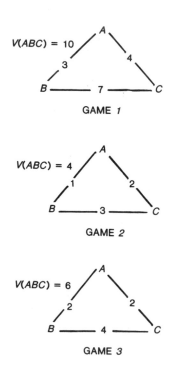

GAME *1*

GAME *2*

GAME *3*

For a more critical discussion of Shapley values, the reader should turn to *Games and Decisions* (1957) by Luce and Raiffa. Two games that I referred to earlier are shown in Figure 6.12, with the Shapley values of each of the players.

For the mathematically sophisticated, this is the way one computes the Shapley value $V(I)$, for an arbitrary player I in an arbitrary *n*-person game:

Figure 6.12

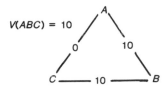

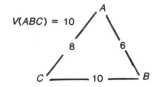

THE SHAPLEY VALUE FOR:
PLAYER *A* IS 5/3
PLAYER *B* IS 20/3
PLAYER *C* IS 5/3

THE SHAPLEY VALUE FOR:
PLAYER *A* IS 7/3
PLAYER *B* IS 10/3
PLAYER *C* IS 13/3

For each coalition S, let $D(S)$ be the difference between the values of the coalition S and the coalition S without player I (if I is not in S, $D(S) = 0$). For each coalition S, *compute* $[(s-1)!(n-s)!/n!] \cdot D(S)$, where s is the number of players in S, n is the number of players in the game, and $n!$ means $n(n-1)\ldots(3)(2)(1)$. Add these numbers for all coalitions S, and the answer is the Shapley value for I.

Shapley's formula may also be derived from a bargaining model. Imagine that at the start one player joins another to form an intermediate two-person coalition, then these two are joined by a third player, and ultimately an *n*-person coalition is formed, one player at a time. Suppose that at each stage the new player gets the marginal gain: the difference in the values of the coalition already formed and the coalition with the new player. If you assume that the final n-person coalition is as likely to form in one way as in another, the expected gain of a player is his or her Shapley value.

A variation of the bargaining model just described was also used to solve a different kind of problem in an unexpected way. S. C. Littlechild and Guillermo Owen (1973), told of one such problem facing airport planners.

A single runway services many different aircraft. The runway's size, and therefore its cost, is determined by its largest user. The planners sought an equitable way of distributing costs among its users to reflect the more modest requirements of smaller aircraft. They adopted the following procedure:

1. The cost of a runway adequate for the smallest user is divided equally among all users.
2. The difference in cost between a runway adequate for the smallest and next smallest aircraft is divided equally among all users *except* for the smallest (which doesn't make use of the additional space).
3. The process is continued taking successively larger aircraft and passing the marginal cost incurred by the larger aircraft on to the larger aircraft until finally the largest user pays all of the marginal costs between it and the next largest user.

So, for example, if the cost of a runway for aircraft A, B, C, and D was 8, 11, 17 and 19, respectively, the costs would be allocated as shown in figure 6.13.

If you imagine that the airplanes are players and the cost of a runway for a group of planes is the value of a coalition, then the cost assigned to each plane is its Shapley value. To calculate airplane B's bill (Shapley value), assume the grand coalition, $ABCD$, forms in all possible ways—twenty-four in all. A quarter of the time B would form first and pay 11; a twelfth of the time AB would form first and B would pay

Figure 6.13

MARGINAL COST

		$\frac{8}{4} = 2$	$\frac{11 - 8}{3} = 1$	$\frac{17 - 11}{2} = 3$	$19 - 17 = 2$	TOTAL COST
	A	2				2
AIRCRAFT	B	2	1			3
	C	2	1	3		6
	D	2	1	3	2	$\frac{8}{19}$

$V(AB) - V(A) = 11 - 8 = 3$; and the rest of the time B would pay nothing because there would be no marginal increase when B joined. B's assessment would therefore be $(1/4)(11) + (1/12)(3) = 3$.

A very similar problem was described by Jeffrey L. Callen (1978): How do you allocate costs among several depreciating assets when these assets are jointly producing revenue? Here again the Shapley value is a reasonable solution.

This technique of allocating costs seems to have an intuitive appeal and has been rediscovered independently in various contexts. Judging from the number of articles in accounting journals, its use is becoming fairly widespread. In some cases intuition proved less reliable, however. Phillip Straffin and J. P. Heaney (1981) describe a similar problem in which a different sort of solution was sought. The Tennessee Valley Authority was created for several purposes—supplying electric power, irrigation, flood control, recreation, and so forth—and the program's costs were to be allocated among them. The planners assumed that any group of services would do better with the dam than they could by spending the same amount without it; that is, they sought an imputation in the core. Because of the great economies introduced by the dam, such a "solution" existed but apparently it occurred to no one that in some cases the core might be empty.

GAME THEORY

The A Priori "Value" of a Player

In March 1868, the directorate of the Erie Railroad, consisting of Daniel Drew, Jay Gould, and James Fisk, sought permission from the New York State Legislature to issue stock at will. They were opposed by Cornelius Vanderbilt of the New York Central Railroad. The general approach required was well understood by the Erie directorate. According to one account (O'Connor 1962), the members of the New York State Legislature "for the most part sold their votes at open bidding in the corridor of the State House." The only real question was how much should be set aside for bribes. If we assume that there are a fixed number of legislators who never abstain from voting, and that permission to issue stock at will has some definite dollar value to the Erie directorate, how much should the Erie directorate pay to bribe one member? Two members? n members? If the bill must be approved by another legislative body and an executive, how will this affect the answers?

Lloyd S. Shapley and Martin Shubik (1954) found one answer based on what amounts to the Shapley value. (You can regard legislative bodies, executives, individual legislators, and so forth, as players in an n-person game; any coalition that has enough votes to pass a bill is called winning, and the others are called losing.) Shapley and Shubik concluded that the power of a coalition was *not* simply proportional to its size; a stockholder with 40 percent of the outstanding stock, for example, would actually have about two-thirds of the power if the other 60 percent of the stock was divided equally among the other six hundred stockholders. (When a person controls 51 percent of the vote, the other 49 percent is worthless if decisions are made by majority vote. Hence, power could not sensibly be defined as being proportional to the number of votes.)

The rationale on which Shapley and Shubik based their evaluation scheme is best illustrated by two examples:

Committee *A* consists of twenty people, nineteen members and chairman. Committee *B* consists of twenty-one members, twenty members and a chairman. In both cases we will assume (for simplicity) that the members never abstain from the vote and that decisions are made by a simple majority. If the members are evenly divided on an issue, the tie-breaking vote is cast by the chairman. What is the power of the chairman in each of the two committees?

In the case of committee *A*, the answer is easy: the chairman has no power at all. The chairman votes only when the members split evenly, and nineteen members never split evenly. (There are no abstentions.) In committee *B*, the situation is a little more complicated. The chairman will cast a vote occasionally but will not vote as often as ordinary members. Does it follow that the chairman, who clearly has some power, is weaker than a member? Shapley and Shubik think not. To see why, imagine that a proposition the chairman favors is before the committee. There are three possibilities: the number of committee members who favor the proposition is (1) less than ten, (2) ten, (3) greater than ten. In case 1, the chairman can't vote and the bill is defeated; but if the chairman could vote, *his vote would be futile.* In case 3, the chairman can't vote but the bill passes anyway; if the chairman could vote, *his vote would be redundant.* Case 2 is the only one in which the chairman can vote; it is also *the only case that matters.* Consequently, Shapley and Shubik conclude, the chairman has the same power as a member.

Using a similar but more general argument, Shapley and Shubik show how an index of power can be derived. They define the power of a coalition (or player) as that fraction (of all possible voting sequences) of the time that the coalition casts the deciding vote; that is, the vote that first guarantees passage. Thus the power of a coalition is always between 0 and 1. A power of 0 means that a coalition has no effect at all on whether a bill is passed; and a power of 1 means that a coalition determines the outcome by its vote. Also, the sum of the powers of all the players is always equal to 1. If there are n players in a game and all votes have the same significance, the power of each player is $1/n$, as one would expect. A simple numerical example will show how these indices are calculated.

Suppose decisions are made by majority rule in a body consisting of A, B, C, and D, who have 3, 2, 1, and 1 vote, respectively. There are twenty-four possible orders for these members to vote, as listed in figure 6.14. For each voting sequence the pivot voter—that voter who first raises the cumulative sum to 4 or more to form a majority—is underlined.

Figure 6.14

A<u>B</u>CD	A<u>B</u>DC	A<u>C</u>BD	A<u>C</u>DB	A<u>D</u>BC	A<u>D</u>CB
B<u>A</u>CD	B<u>A</u>DC	BC<u>A</u>D	BC<u>D</u>A	BD<u>A</u>C	BD<u>C</u>A
C<u>A</u>BD	C<u>A</u>DB	CB<u>A</u>D	CB<u>D</u>A	CD<u>A</u>B	CD<u>B</u>A
D<u>A</u>BC	D<u>A</u>CB	DB<u>A</u>C	DB<u>C</u>A	DC<u>A</u>B	DC<u>B</u>A

By counting, you find that A is pivotal in twelve of the twenty-four sequences while each of the other members is pivotal in four; therefore A has an index of power one-half and the others have an index of power one-sixth. Curiously, although B has twice as many votes as C (and D), B has no more power. When you consider that A's vote determines the outcome unless the others unite against A, it becomes clear that B, C, and D play identical roles in the voting, the different weights notwithstanding. This is reflected in the power indices.

Some Applications of the Shapley-Shubik
Index of Power

The Shapley-Shubik index—an abstract measure of power—seems far removed from living and breathing decision-making bodies. But the numbers it generates are often suggestive of power relationships in the real world. Consider a few examples:

Suppose that in a voting body with $2n+1$ members in which decisions are made by majority rule, a single strong member has k votes and the remaining $(2n-k+1)$ members have one vote. It then turns out that the power of the strong member is $k/(2n+2-k)$. As k increases, the power of the strong member increases disproportionately until it approaches half the total vote and the strong member gains virtually all the power. On the other hand, if there are two "strong" members each with n votes and a "weak" member with a single vote, everyone has the same power, and the "weak" member has more influence than the size of his or her vote would suggest.

The implications of these simple observations may be seen at political conventions where delegates are pulled in two directions. Support blocs form for the various candidates, and these behave like a single candidate with many votes. If the ultimate winner follows a clear path to victory, the large rewards go to early supporters who went out on a limb. But if delegates commit themselves early, they may back the wrong horse or miss out on an opportunity of being the single vote that the two large blocs must woo. At a certain point the frontrunner is perceived as the likely winner and a *bandwagon* starts: the bloc is joined by a wave of uncommitted voters and there are wholesale defections from the other blocs.

On theoretical grounds Riker and Brams formulated a theory for bandwagon formation. Straffin, in a delightful mixture of theory and practice, applied the theory to the 1976 Republican primary. At various times during the campaign it was asserted that a bandwagon was underway, but the mathematical model indicated otherwise. When the mathematical model finally signaled a bandwagon for Gerald Ford, events seemed to confirm this conclusion. Uncharacteristically, Ronald Reagan committed himself to a liberal runningmate, and Reagan's campaign manager reported that his candidate had a larger number of committed delegates than he really did, indications that Reagan perceived desperate measures were in order because a bandwagon had been formed.

Blocs are formed in many voting bodies; they are not restricted to political conventions. Members retire to a back room, take a position after a vote among themselves, and vote unanimously on that position when they return to the parent body; they act like a single player with many votes.

Forming a bloc often has the effect of increasing your power, but not always; if there are five voters with voting weights (2, 2, 1, 1, 1), the power of a player with weight 1 is two-fifteenths, so the combined power of all the weight-1 players is two-fifths. If the three 1's form a bloc, the weights of the players become (2, 2, 3) and each voter has a power index of one-third. The combined power of the original 1 voters drops from two-fifths to one-third after the coalition is formed.

Now see what happens if two of five voters combine to change the vote structure from (1, 1, 1, 1, 1) to (2, 1, 1, 1); the immediate effect on the two is to increase their power from two-fifths to one-half. But this may provoke two other single voters to join in retaliation so that the effective weights of the voters become (2, 2, 1). Now *both* blocs have a combined power of one-third, which is less than the two-fifths they started with.

214

Phillip Straffin (1977) described just such a situation. The two largest cities in Rock County, Wisconsin—Janesville and Beloit—had fourteen and eleven representatives, respectively. For many issues the two cities had opposing interests, and there were many suggestions that the representatives of Beloit would do better to vote as a bloc. Nevertheless, the representatives from both cities continued to vote independently. Although bloc voting would have increased their power in the short run, Straffin felt that the representatives anticipated retaliation that would lower the power of both large cities (just as it did in the preceding example).

Power indices—a variation of those used by Shapley and Shubik—were the basis of a successful suit by John Banzhaf (1965) against a weighted voting system that was used by Nassau County, New York, in 1958. Six municipalities had (9, 9, 7, 3, 1, 1) representatives in the Board of Supervisors; it takes only a simple calculation to confirm that the three smallest towns were disenfranchised.

The Shapley-Shubik indices have been applied to various voting bodies, and sometimes the power distributions they reflect are surprising. In the United Nations Security Council, for example, in which all five permanent members and four of the ten rotating members must agree for a measure to pass, the five permanent members control 98 percent of the power. The index may be applied where there are several legislative bodies acting together as well. It takes a majority of the House of Representatives and Senate to pass a bill with the President's approval and two-thirds of each of these bodies to pass one without it. The House and Senate as a body each have about five-twelfths of the power (individual senators are more powerful than members of the House since there are fewer of them), and the President has the remaining one-sixth.

GAME THEORY

From Individual Preferences to Social Choice

A group of individuals may at times be called on to make a single, joint decision. Families, legislative bodies, stockholders, national electorates, committees, juries, the Supreme Court of the United States are all in this position. The decision of the group is dependent on the decisions of the membership, but not always in the same way. Everyone may be allowed a single vote; or the votes may be weighted to reflect the wealth of the voters or the number of shares owned by each one. There may be an absolute veto such as there is, under certain conditions, at the United Nations, or a partial veto such as is available to the President of the United States or to the House of Lords in England.

The conversion of individual wants into societal decisions seems straightforward, but it is really very involved. I will only discuss the simplest case—where everyone's opinion is equally important—but even in this situation what appear to be equitable decision rules turn out to have paradoxical consequences.

There are two different questions you can ask about voting games: (1) How should you convert the preferences of group members into a single group decision (when choosing a mayor or town council, for example)? and (2) Once you have established a voting mechanism to translate these preferences, how should an individual member vote to best attain his or her own ends?

These questions are related, and both have been around for some time. One attempt to derive the rules governing strategic voting—the answer to question 2—was made in the middle of the last century. At that time a constituency in England was assigned a fixed number of seats in Parliament and each voter could cast a fixed number of votes. Political parties submitted a slate of any number of candidates, and the party faithful could support them all equally or favor some over others. In a paper written in 1853, James Garth Marshall used what was essentially a minimax argument to calculate the optimal number of candidates a party should field. This optimal number was always at least as great as the number of votes a voter could cast and never greater than the number of seats assigned to the constituency.

About thirty years later, Lewis Carroll, the celebrated author, addressed the problem discussed in question 1. Assuming parties would act optimally (as Marshall had suggested earlier), Carroll derived the appropriate number of votes that should be assigned each voter and the appropriate number of seats that should be assigned each constituency to make government most representative (in a sense he defined in his paper).

Since it may not be clear why there should be *any* difficulty in devising rules for making social choices, let's consider some "reasonable" possible rules and see where they go wrong.

Let's suppose that a group of 300 people must choose one of three alternatives: *A, B,* or *C.* Why not choose an alternative that a majority prefers? Suppose $(ABC) = 101$, $(BCA) = 100$, and $(CBA) = 99$. (I use "$(BCA) = 100$" to indicate there are 100 people who prefer *B* most and *A* least). In such a case there is *no* single alternative that a majority prefers. This suggests another possibility, however—choose the alternative that a plurality prefers. Since 101 people prefer *A,* this would be the group's choice according to the rule, but the decision might be questioned on other grounds. For one thing, almost two-thirds of the voters consider *A* the *worst* choice; for another, in a two-way race between either *A* and *B* or *A* and *C,* *A* would be defeated by almost a two-to-one margin.

This last situation is not just of academic interest, incidentally. In the 1970 New York senatorial elections two liberals,

Richard L. Ottinger and Charles E. Goodell, and one conservative, James Buckley, were competing. Their respective percentage of the total vote in the election was 37, 24, and 39. Since Ottinger and Goodell had similar positions, it was generally accepted that the supporters of each would prefer the other liberal candidate to Buckley. It appears that plurality voting permitted Buckley to win an election in which he was least favored by 61 percent of the vote.

In the 1969 New York City mayoralty election the same situation arose again, but here the roles of conservatives and liberals were reversed; the liberal John Lindsay defeated the more conservative John Marchi and Mario Biaggi.

One way of avoiding this problem—a method adopted in France and in some state elections—is to have more than one vote; in the earlier vote(s) the weakest alternatives are eliminated, and in the final vote the best of the two remaining alternatives is selected by majority vote. If this method were adopted, Goodell would have been eliminated in the first election and his supporters, who would have presumably switched to Ottinger, would have helped elect a liberal; in the second election a conservative would have been elected for the same reason.

But now consider what happens when $(BAC) = 101$, $(CAB) = 101$, and $(ABC) = 98$; on the first ballot A is eliminated, and in the second round B would eliminate C by almost two to one. And this despite the fact that A would win a two-way contest against either B or C by almost two to one.

To avoid this last situation, there is another possibility—vote on the alternatives two at a time, pitting the winners of each vote against one another until only the winning alternative remains. But now suppose $(ABC) = (BCA) = (CAB) = 100$, and suppose further that A and B are paired in the first election. A wins the first vote but loses the second one to C.

But *C*'s victory is not convincing since it depends on the order in which the contestants were paired (this should be clear from symmetry). Whatever the voting order, the alternative that doesn't compete in the first round will ultimately be chosen.

Another curious thing about all of this is that the preferences of society are what a mathematician would call intransitive; if three two-way votes were taken, society would prefer *A* to *B*, *B* to *C*, and *C* to *A*. It is as though a person preferred pie to custard, custard to ice cream, and ice cream to pie. Such a person would be hard to satisfy. It is not too surprising that fashioning group decisions—which often have intransitive preferences—is a tricky business.

And it even gets worse! In the last example (with the same voting order), it should be clear to those with preference order (*ABC*) that their worst preference order will be picked unless they do something about it. By shifting their vote from *A* to *B* they insure *B*'s victory in both the first and second votes. While it might seem that others might disguise their true preferences and change the outcome once again, they won't; those with preference order (*BCA*) will be even more pleased with the new outcome, and voters with a (*CAB*) preference order will be frustrated but helpless.

Strategic voting—voting in a way that does not express your true preferences—is commonplace in practical politics. In his book *Paradoxes in Politics* (1976), Steven J. Brams mentions several examples of elections where it was applied or where it could have been.

In the 1948 presidential elections, the polls gave Henry Wallace, the Progressive Party candidate, more votes than he actually received in the election. Many political commentators felt that a substantial number of Wallace supporters defected to the ultimate winner, Harry Truman, to avoid the election of a Republican. And in 1968 the general feeling was that the supporters of George Wallace, another third-party candidate, supported Richard Nixon to avoid the election of a Democrat. In 1912, on the other hand, the Democrats who received 42 percent of the vote won the election because the Progressive and Republican voters—both of whom presumably wanted the Democratic candidate, Wilson, least—failed to vote strategically. Perhaps this was because both parties had about the same strength and neither was willing to concede it was backing a lost cause.

In 1956 a school construction bill was submitted to the House of Representatives; an amendment to it was proposed to cut off aid to segregated schools. Northern Democrats wanted an amended bill most but preferred the original bill to none at all. Southern Democrats wanted the original bill most but preferred no bill to an amended one. Republicans wanted no bill most but preferred the amended bill to the original one. The House made its decision in two stages: the amendment was considered first and then a vote was taken on the amended, or unamended, bill. The Northern Democrats, who voted sincerely for the amendment rather that strategically against it, were outnumbered by the Republicans and Southern Democrats in the final vote. In a similar situation in the Senate in 1955, the Northern Democrats settled for their second-best preference—a construction bill without the amendment—and this unamended bill passed with the help of the Southern Democrats.

A common practice in most legislative bodies, and another example of strategic voting, is logrolling: an agreement by two or more members to support each other's pet bills. Such an arrangement must be beneficial to everyone, it would seem, since losers would not participate. But Eric M. Uslaner and J. Ronnie Davis (1975) constructed an example in which every party came out behind. Three towns—A, B, and C— were voting on six bills—X, Y, Z, U, V, W. The gains/losses

Figure 6.15

		BILLS					
		X	Y	Z	U	V	W
	A	3	3	2	−4	−4	2
TOWNS	B	2	−4	−4	2	3	3
	C	−4	2	3	3	2	−4

for each town that would result in the passage of each bill are shown in figure 6.15.

If everyone votes sincerely, each bill passes and each town gains 2. But if A votes against Z in return for B's vote against U, each will gain 2. Similar agreements by B and C, and A and C, concerning bills X and Y, and V and W, respectively, will have the same result. Finally, after these three "profitable" agreements each town loses the 2 it would have had if it had voted sincerely.

The paradox is easily resolved: the same mechanism that we observed in the prisoner's dilemma is operating here. In the prisoner's dilemma, players, following their own apparent interests, gained but inflicted greater losses on their partners. The net effect of everyone playing uncooperatively was a loss for everyone. Here each partnership agreement yields a total of 4 to the insiders but costs the outsider 6; when everyone participates in such agreements, the result is predictable.

Arrow's Theorem

The solution to the problem raised in the last section—a "sensible" decision rule that translates every possible pattern of membership preferences into a single set of societal preferences—seems very elusive, judging by my lack of success and the few examples I gave from practical politics. But this lack

221

of success may not reflect a lack of cleverness; it may be intrinsic to the problem.

Kenneth Arrow (1951) decided not to look for a reasonable decision rule but instead tried to formulate general principles that *any* reasonable decision rule would satisfy.

Arrow assumed that a voting body was faced with at least three alternatives and that members' preferences were transitive—if A is better than B, and B better than C, A must necessarily be better than C. He assumed no single member would dictate the preferences of society. If, for some membership preference pattern, A was the preferred alternative, then it would continue to be so if A gained additional support. Also, no alternative was to be ruled out in advance: each alternative must have *some* preference pattern that would induce society to adopt it. And Arrow required one thing more—if a membership preference pattern induced a societal preference ordering of some subgroup of alternatives and some members changed their preferences but only with respect to alternatives outside of that subgroup, the original societal pattern of alternatives within the subgroup would remain undisturbed.

Although each of these conditions seems reasonable and most of them are compelling, Arrow proved that *no decision rule can satisfy them all*. So if you adopt Arrow's view, any decision rule must make unreasonable decisions at least some of the time.

The Alabama Paradox

A final example of how a "simple" solution to a voting problem has caused considerable confusion in this country is known as the Alabama Paradox. The number of members in the House of Representatives from each state must be proportional to its population, according to the United States Constitution. Apportioning these representatives to the states seems a trivial task, but one such method, suggested by Alexander Hamilton, turned into a time bomb. The Alabama Paradox described by Steven J. Brams (and based on a 1975 paper by Michael L. Balinski and H. P. Young) again illustrates the danger of "common sense" solutions

The basic problem is simple enough; once the size of the House of Representatives is determined, you can easily calculate the *ideal* number of representatives from each state. But this ideal number will generally contain a fraction, and a state must have a whole number of representatives. The bits and pieces of fractional representatives must be sorted out so that each state has an integral number of representatives— some slightly overrepresented and some slightly underrepresented. Hamilton's method seemed to be a reasonable and equitable approach.

Hamilton's plan was implemented in two steps. First, each state's exact representation in the House was calculated and expressed as a whole number plus a proper fraction. Each state then received its whole number of representatives; an additional representative was allocated to those states with the highest fractional values to bring the total number of representatives to its predetermined value: the size of the House of Representatives. A simple example will make the procedure clear:

Assume there are five states with populations of 100, 150, 200, 250, and 300, and the size of the House is 19. (See figure 6.16.)

The ideal representation is obtained by multiplying the size of the House by the fraction of the total population contained in the state; the minimum representation is the integral part of the ideal representation. Since there is a gap of

Figure 6.16

POPULATION	IDEAL REPRESENTATION	MINIMUM REPRESENTATION	ADDITIONAL MEMBER	ACTUAL TOTAL REPRESENTATION
100	1.9	1	1	2
150	2.85	2	1	3
200	3.8	3	1	4
250	4.75	4	1	5
300	5.7	5	0	5
1,000	19.0	15	4	19

four between the nineteen members of the House and the total minimum representation, states with the four highest fractions are awarded an additional representative, yielding the actual total representation.

Hamilton's method was adopted in 1850 and dropped in 1900; before it was dropped it caused considerable furor. The reason will be clear if you work out the following simple example:

Imagine three states—A, B, and C—with 380, 380, and 240 people, respectively. Now consider two cases: the representative body contains (1) 14 seats and (2) 15 seats. (See figure 6.17.)

Although the House size increased from 14 to 15 and the population remained unchanged, state C *lost* representation in the House!

After the 1900 census the number of representatives from each state was calculated for House sizes from 350 to 400. It

Figure 6.17

STATE	POPULATION	CASE 1 IDEAL REPRESENTATION	CASE 1 TOTAL REPRESENTATION	CASE 2 IDEAL REPRESENTATION	CASE 2 TOTAL REPRESENTATION
A	380	5.32	5	5.7	6
B	380	5.32	5	5.7	6
C	240	3.36	4	3.6	3
TOTAL	1,000	14.0	14	15.0	15

turned out that Colorado had three representatives whatever the House size except when it was exactly 357; if there were 357 members, it only had two representatives. Oddly enough, it was precisely a House size of 357 that the majority coalition wanted. The Hamilton apportionment was denounced as a "freak" and an "atrocity" by members whose seats were in danger, and the matter was finally resolved by fixing the number at 386; a size that enabled every state to hold on to the seats that it held during the previous session. The Alabama Paradox arose because a "reasonable solution" to a problem was adopted without sufficient analysis. What appear to be "common sense" answers often turn out to be "freaks" and "atrocities," especially in political science.

Solutions to Problems

1. a. It is not clear whether all three should or will join together; it depends on a number of factors, such as the persuasiveness, aspirations, and utilities of the players. The Aumann-Maschler theory prescribes that the engineer get 16/3, the lawyer get 28/3, and the treasurer get 40/3 if all three join; according to Shapley, the respective numbers are 22/3, 28/3, and 34/3. What is clear is that each pair will *not* get as much as each person could have acting alone (if they did, their total payoff would have to be at least 30). No matter what the payoff, some pair can do better; that is, the core is empty. Nevertheless, all three may join together for the extra security because if a pair does form, some player must be left out.

b. If a pair forms, the engineer will get 6, the lawyer, 10, and the treasurer, 14, as long as each is a member; according to Aumann and Maschler, a player receives the same amount whatever coalition he or she joins.

c. The Aumann-Maschler theory predicts that a three-person coalition must have a solution in the core—no pair can do better by leaving. The payoffs in the core may vary considerably however; (0, 16, 24), (16, 0, 24), and (16, 20, 4) are all possible payoffs for the engineer, lawyer, and treasurer, respectively. The Shapley payoffs are 34/3, 40/3, and 46/3, respectively. The von Neumann-Morgenstern theory allows for many more possibilities.

2.   Taking the Shapley value approach described in the text (not the only solution but a reasonable one), the coalition values would be: $V(A) = 6,000$, $V(AB) = V(B) = 8,000$, and $V(C) = V(AC) = V(BC) = V(ABC) = 10,000$. A, B, and C should pay $2,000, $3,000, and $5,000, respectively.

3. a.   It appears that the chairman *must* be the most powerful by any rational standard since he has all the options of the others and more—he has the important advantage of breaking ties when a majority can't be formed. If you accept this, try to find the error in the following chain of reasoning:

(I) C *will vote for* Y.

If A and B vote the same way, it doesn't matter what C does; if they don't, C's vote will be decisive and C should naturally choose the alternative he or she prefers the most.

(II) A *will vote for* X.

Assuming C votes for his or her most preferred alternative, Y, the outcome will depend on the votes of A and B as shown in figure 6.18:

It is obvious from this matrix that A's strategy X dominates Y and Z.

(III) B *will vote for* X.

If B accepts steps (I) and (II), he or she will vote for X since B prefers X to Y. The final outcome, then, must be X, *the alterna-*

Figure 6.18

		B'S VOTE		
		X	Y	Z
	X	X	Y	Y
A'S VOTE	Y	Y	Y	Y
	Z	Y	Y	Z

tive preferred least by the chairman. (This example is due to Steven J. Brams.)

b. Using the Shapley value as a measure, the power of the smaller borough presidents increased substantially from 3/56 to 3/35, the power of the larger borough presidents decreased from 1/8 to 3/35, and the other three members lost, but just barely— they dropped from 11/56 to 4/21.

c. (i) If we use the Shapley value as our measure, we find voters often gain by forming a bloc. If there are five voters, each with a single vote, and decisions are made by majority rule, two players can increase their joint power from two-fifths to one-half by acting as a unit. On the other hand, in a body with two two-vote members and three single-vote members, the single-vote members decrease their power from two-fifths to one-third by joining together.

(ii) It is possible for an old member of a voting body to gain power when a new member is added. If there are three voters with thirteen, seven, and seven votes, each has one third of the power. By adding a three-vote member, the thirteen-vote member gets half the power. In each case that member needs one more member to form a majority. It is clearly easier to do this in the second case, since there are more possible members who can join the thirteen-vote member. If there is one five-vote member and four single-vote members originally, adding a new two-vote member gives the weak voters some power where they had none before.

d. It is generally accepted that the large states have more power than the small ones. This is so even after you allow for the two-vote senatorial bonus.

e. Call the one-vote and two-vote members type A, the three-vote and four-vote members type B; and the five-vote member type C; to win an election you need (i) one type-C and one type-B member, (ii) one type-C member and two type-A members, or (iii) two type-B members and one type-A member. When expressed this way, it becomes clear that the two type-A members and the two type-B members are indistinguishable. Both A members have Shapley values of 2/30, both B members have Shapley values of 7/30, and the C member has a Shapley value of 12/30.

4. Depending on the circumstances, logrolling can either be good or bad for society as a whole. Uslaner's example in the text shows that logrolling can be detrimental to society, but if every "−4" payoff is changed to a "−40" payoff, the same example shows that vote trading can be socially beneficial.

5. a. Although individuals' preferences may be transitive, the preferences of groups often are not, whatever the voting procedure.

b. Individuals often do best if they don't vote their true preferences, and plurality voting has its own difficulties, as we saw in the text.

c. It may be that *every alternative* is less preferred than some other alternative; in such a case this condition could never be satisfied, whatever the voting procedure.

6. Legislators who were threatened with the loss of their seats in the House when the size of the House was *increased* saw good reason to object to the plan.

7. If there is a majority preference, (a) is a reasonable plan; but with three or more options, it is possible that there is none.

Both (b) and (c) are plausible but each leads to an outcome that, under certain circumstances, many would consider inappropriate. A more detailed description of the possible difficulties is found in the text.

Bibliography

Allais, Maurice. "Le Comportement de l'Homme Rationnel devant le Risque: Critiques des Postulates et Axiomes de l'Ecole Americaine." *Econometrica*, 21 (1953):503–546.

Allen, Layman E. "Games Bargaining: A Proposed Application of the Theory of Games to Collective Bargaining." *Yale Law Journal* 165 (1956):660–693.

Ankeny, Nesmith C. *Poker Strategy—Winning with Game Theory.* New York: Basic Books, 1981.

Anscombe, F. J. "Applications of Statistical Methodology to Arms Control and Disarmament." In *Final Report to the U.S. Arms Control and Disarmament Agency under Contract No. ACDA/ST-3.* Princeton, N.J.: Mathematica Inc., 1963.

Arkoff, A., and Vinacke, W. E. "An Experimental Study of Coalitions in the Triad." *American Sociological Review* 22 (1957):406–414.

Arrow, K. J. *Social Choice and Individual Values.* Cowles Commission Monograph 12. New York: John Wiley and Sons, Inc., 1951.

Aumann, R. J., and Maschler, Michael. "The Bargaining Set for Cooperative Games." In *Advances in Game Theory*, Annals of Mathematics Study 52, edited by M. Dreshner, L. S. Shapley, and A. W. Tucker, pp. 443–476. Princeton: Princeton University Press, 1964.

———. "Some Thoughts on the Minimax Principle." *Management Science* 18 (1972):50–54.

Aumann, R. J., and Peleg, B. "Von Neumann-Morgenstern Solutions to Cooperative Games without Side Payments." *Bulletin of the American Mathematical Society*, 66 (1960):173–179.

Avenhaus, R., and Frick, H. "Game Theoretical Treatment of Material Accountability Problems." *International Journal of Game Theory* 5 (1976):41–49; 6 (1977):117–135.

Axelrod, R. "Effective Choice in the Prisoner's Dilemma." *Journal of Conflict Resolution* 24 (1980a):3–25.

———. "More Effective Choice in the Prisoner's Dilemma." *Journal of Conflict Resolution* 24 (1980b):379–403.

———. "The Emergence of Cooperation among Egoists." *American Political Science Review* 75 (1981):306–318.

———, and Hamilton, W. D. "The Evolution of Cooperation." *Science*, 211 (1981):1390–1396.

Balinski, M. L., and Young, H. P. "A New Method for Congressional Apportionment." *American Mathematical Monthly* 82 (1975):701–730.

Banzhaf, J. F., III. "Weighted Voting Doesn't Work: A Mathematical Analysis." *Rutgers Law Review* 19 (1965):317–343.

Bartoszynski, R., and Puri, M. "Some Remarks on Strategy in Playing Tennis." *Behavioral Science* 26 (1981):379–387.

Becker, Gordon M., and De Groot, Morris H. "Stochastic Models of Choice Behavior." *Behavioral Science* 8 (1963):41–55.

———, and Marchak, Jacob. "An Experimental Study of Some Stochastic Models for Wagers." *Behavioral Science* 8 (1963):199–202.

Berkovitz, L. D., and Dresher, Melvin. "A Game-Theory Analysis of Tactical Air War." *Operations Research* 7 (1959):599–620.

———. "Allocation of Two Types of Aircraft in Tactical Air War: A Game-Theoretic Analysis." *Operations Research* 8 (1960):694-706.

Billera, Louis J. "Some Results in n-Person Game Theory." *Mathematical Programming* 1 (1971):58-67.

Bixenstine, V.; Gabelin, Edwin; and Gabelein, Jacquelyn W. "Strategies of 'Real' Others Eliciting Cooperative Choice in a Prisoner's Dilemma." *Journal of Conflict Resolution* 15, 2 (1971):157–166.

Bixenstine, V.; Polash, Herbert M.; and Wilson, Kellogg V. "Effects of Level of Cooperative Choice by the Other Player on Choices in a Prisoner's Dilemma Game." *Journal of Abnormal and Social Psychology* 66 (1963):308–313 (part 1); 67 (1963):139–147 (part 2).

Black, Duncan. "The Decision of a Committee Using a Special Majority." *Econometrica* 16 (1948):245–261.

———. *The Theory of Committees and Elections.* Cambridge, England: Cambridge University Press, 1958.

———. "Lewis Carroll and the Theory of Games." *American Economic Review* 159 (1969):206–216.

Blaquire, Austin; Gérard, Françoise; and Leitman, George. "Quantitative and Qualitative Games." *Mathematics in Science and Engineering* 58 (1969).

Bond, John R., and Vinacke, Edgar W. "Coalitions in Mixed-Sex Triads." *Sociometry* 24 (1961):61–81.

Bonoma, Thomas V.; Tedeschi, J. T.; and Linskold, S. "A Note Regarding an Expected Value Model of Social Power." *Behavioral Science* 17 (1972):221–228.

Brams, Steven J. *Paradoxes in Politics.* New York: Free Press, 1976.

———, and Davis, Morton D. "Resource-Allocation Models in Presidential Campaigns: Implications for Democratic Representation." *Annals of the New York Academy of Sciences* 219 (1973):105–123.

———. "The 3/2's Rule in Presidential Campaigning." *American Political Science Review* 68 (1974):113–134.

———. "Optimal Jury Selection: A Game-Theoretic Model for the Exercise of Peremptory Challenges." *Operations Research* 26 (1978):966–991.

———, and Straffin, P. D. "The Geometry of the Arms Race." *International Studies Quarterly* 23 (1979):567–588.

Brams, Steven J., and Riker, W. H. "Models of Coalition Formation in Voting Bodies." *Mathematical Applications of Political Science*, vol. 6, edited by James F. Herndon and Joseph L. Bernd, pp. 79–124. Charlottesville: University of Virginia Press.

Brayer, Richard. "An Experimental Analysis of Some Variables of Minimax Theory." *Behavioral Science* 9 (1964):33–44.

Buchanan, James M. "Simple Majority Voting, Game Theory and Resource Use." *Canadian Journal of Economic and Political Science* 27 (1961):337–348.

———, and Tullock, Gordon. *The Calculus of Consent.* Ann Arbor, Mich.: University of Michigan Press, 1962.

Callen, Jeffrey L. "Financial Cost Allocations: A Game-Theoretic Approach." *Accounting Review* 53 (1978):303–308.

Caplow, Theodore. "A Theory of Coalition in the Triad." *American Sociological Review* 21 (1956):489–493.

———. "Further Developments of a Theory of Coalitions in the Triad." *American Journal of Sociology* 66 (1959):488–493.

Carroll, Lewis. "The Principles of Parliamentary Representation," 1st ed. November 1884 (booklet).

Case, James. "A Different Game in Economics." *Management Science* 17 (1970/71):394–410.

Cassady, Ralph, Jr. "Taxicab Rate War: Counterpart of International Conflict." *Journal of Conflict Resolution* 1 (1957):364–368.

——— . "Price Warfare in Business Competition: A Study of Abnormal Competitive Behavior." Occasional Paper No. 11. The Graduate School of Business Administration, Michigan State University. N.D.

Caywood, T. E., and Thomas, C. J. "Applications of Game Theory in Fighter Versus Bomber Conflict." *Operations Research Society of America* 3 (1955):402-411.

Chacko, George K. *International Trade Aspects of Indian Burlap.* New York: Bookman Associates, Div. of Twayne Publishers, 1961.

——— . "Bargaining Strategy in a Production and Distribution Problem." *Operations Research* 9 (1961):811-827.

Chaney, Marilyn V., and Vinacke, Edgar. "Achievement and Nurturance in Triads in Varying Power Distributions." *Journal of Abnormal and Social Psychology* 60 (1960):175-181.

Chidambaram, T. S. "Game Theoretic Analysis of a Problem of Government of People." *Management Science* 16 (1970):542-559.

Coombs, C. H., and Pruitt, D. G. "Components of Risk in Decision Making: Probability and Variance Preferences." *Journal of Experimental Psychology* 60 (1960):265-277.

Crane, Robert C. "The Place of Scientific Techniques in Mergers and Acquisitions." *The Controller* 29 (1961):326-342.

Dahl, Robert A. "The Concept of Power." *Behavioral Science* 22 (1957):201-215.

Davis, John Marcell. "The Transitivity of Preferences." *Behavioral Science* 3 (1958):26-33.

Davis, Morton. "A Bargaining Procedure Leading to the Shapley Value." Research Memorandum No. 61. Princeton, N.J.: Econometric Research Program, Princeton University, 1963.

——— . "Some Further Thoughts on the Minimax Theorem." *Management Science* 20 (1974):1305-1310.

——— , and Maschler, Michael. "Existence of Stable Payoff Configurations for Cooperative Games." *Bulletin of the American Mathematical Society* 69 (1963):106-108.

Day, Ralph L., and Kuehn, Alfred. "Strategy of Product Quality." *Harvard Business Review* 40 (1962):100-110.

Deutsch, Karl W. "Game Theory and Politics: Some Problems of Application." *Canadian Journal of Economics and Political Science* 120 (1954):76-83.

Deutsch, Morton. "The Effect of Motivational Orientation upon Trust and Suspicion." *Human Relations* 13 (1960a):123-139.

——— . "Trust, Trustworthiness, and the F-Scale." *Journal of Abnormal and Social Psychology* 61 (1960b):366-368.

——— . "The Face of Bargaining." *Operations Research,* 19 (1961):886-897.

Dostoevsky, Fyodor. *The Gambler and Poor Folk*. New York: Everyman's Library, 1915.

Downs, Anthony. *An Economic Theory of Democracy*. New York: Harper & Brothers, 1957.

———. "Why the Government Budget Is Too Small in a Democracy." *World Politics* 12 (1960):541–563.

Dresher, Melvin. *Games of Strategy: Theory and Applications*. Englewood Cliffs, N.J.: Prentice-Hall, 1961.

Dumett, Michael, and Farquharson, Robin. "Stability in Voting." *Econometrica* 29 (1961):33–43.

Edwards, Ward. "Probability-Preference in Gambling." *American Journal of Psychology* 66 (1953):349–364.

———. "The Theory of Decision Making." *Psychological Bulletin* 51 (1954a):380–417.

———. "Probability-Preference among Bets with Different Expected Values." *American Journal of Psychology* 67 (1954b):56–67.

———. "The Reliability of Probability Preferences." *American Journal of Psychology* 67 (1954c):68–95.

———. "Variance Preferences in Gambling." *American Journal of Psychology* 67 (1954d):441–452.

Farquharson, Robin. *Theory of Voting* (New Haven, Conn.: Yale University Press, 1969).

Fellner, William. *Competition among the Few*. New York: Knopf, 1949.

Flood, Merrill M. "Some Experimental Games." Rand Memorandum RM-789-1, 1952.

———. "Some Experimental Games." *Management Science* 5 (1958):5–26.

Forst, Brian, and Lucianovic, Judith. "The Prisoner's Dilemma: Theory and Reality." *Journal of Criminal Justice* 5 (1977):55–64.

Fox, John. "The Learning of Strategies in a Simple Two-Person Game Without a Saddle Point." *Behavioral Science* 17 (1972):300–308.

Fouraker, Lawrence E., and Siegel, Sidney. *Bargaining and Group Decision Making*. New York: McGraw-Hill, 1960.

———. *Bargaining Behavior*. New York: McGraw-Hill, 1963.

Friedman, Lawrence. "Game-Theory Models in the Allocations of Advertising Expenditures." *Operations Research* 6 (1958):699–709.

Galbraith, John Kenneth. *American Capitalism—The Concept of Countervailing Power*. Boston: Houghton Mifflin Co., 1952.

Gale, David, and Stewart, F. M. "Infinite Games of Perfect Information." In *Contribution to the Theory of Games*, edited by H. W. Kuhn and A. W. Tucker, pp. 245–266. Princeton: Princeton University Press, 1953.

Bibliography

Gamson, William A. "A Theory of Coalition Formation. *American Sociological Review* 26 (1961):373–382.

——— . "An Experimental Test of a Theory of Coalition Formation." *American Sociological Review* 26 (1961):565–573.

Gately, D. "Sharing the Gains from Regional Cooperation: A Game-Theoretical Application to Planning Investment in Electric Power." *International Journal of Game Theory* 15 (1974):195–208.

Goehring, Dwight J., and Kahan, James P. "Responsiveness in Two-Person, Zero-Sum Games." *Behavioral Science* 18 (1973):27–33.

Gold, V. "Spieltheorie und Politische Realität." *Politische Studie* (Munich), 191 (1970):257–277.

Griesmer, James H., and Shubik, Martin. "Toward a Study of Bidding Processes: Some Constant-Sum Games." *Naval Logistics Research Quarterly* 10 (1963a):11–22.

——— . "Toward a Study of Bidding Processes, Part II: Games with Capacity Limitations." *Naval Logistics Research Quarterly* 10 (1963b):151–173.

——— . "Toward a Study of Bidding Processes, Part III: Some Special Models." *Naval Logistics Research Quarterly* 10 (1963c):199–217.

Griffith, R. M. "Odds-Adjustment by American Horse-Race Bettors." *American Journal of Psychology* 62 (1949):290–294.

Guyer, Melvin J., and Rapoport, Anatol. "2X2 Games Played Once." *Journal of Conflict Resolution* 16 (1972):409–431.

Guyer, Melvin, and Perkel, B. "Experimental Games: A Bibliography." Ann Arbor, Mental Health Research Institute, Univ. of Michigan, Communication 293, 1972.

Haigh, John, and Rose, Michael. "Evolutionary Game Auctions." *Journal of Theoretical Biology* 85 (1980):381–397.

Hamburger, Henry. *Games as Models of Social Phenomena.* San Francisco: W. H. Freeman and Co., 1979.

Hamilton, John A. "The Ox-Cart Way We Pick a Space-Age President." *New York Times Magazine,* October 20, 1968, p. 36.

Hamilton, William D. "The Genetical Evolution of Social Behavior." *Journal of Theoretical Biology* 7 (1964):1–52.

Hamlen, S.; Hamlen, W.; and Tschikhart, J. "The Use of Core Theory in Evaluating Joint Cost Allocation Schemes." *International Economic Review* 52 (1977):216–267.

Harsanyi, John C. "Approaches to the Bargaining Problem Before and After the Theory of Games: A Critical Discussion of Zethuen's, Hicks', and Nash's Theories." *Econometrica* 24 (1956):144–157.

——— . "Bargaining in Ignorance of the Opponent's Utility Function." *Journal of Conflict Resolution* 6 (1962):29–38.

———. "A Bargaining Model for the Cooperative n-Person Game." In *Contribution to the Theory of Games,* edited by A. W. Tucker and R. D. Luce, pp. 325–355. Princeton: Princeton University Press, 1959.

———, and Selten, Reinhard. "A Generalized Nash Solution for Two-Person Bargaining Games with Incomplete Information." *Management Science* 18 (1972):80–106.

Haywood, O. G., Jr. "Military Decisions and Game Theory." *Journal of the Operations Research Society of America* 2 (1954):365–385.

Hobbes, Thomas. *The Leviathan,* edited by Michael Oakeshott. New York: Collier, 1962. (Originally published 1651.)

Hoffman, Paul T.; Festinger, Leon; and Douglas, Lawrence H. "Tendencies Toward Group Comparability in Competitive Bargaining." In *Decision Processes,* edited by R. M. Thrall, C. H. Coombs, and R. L. Davis, pp. 231–253. New York: John Wiley and Sons, 1954.

Hotelling, Harold. "Stability in Competition." *Economics Journal* 39 (1929):41–57.

Iklé, Charles, and Leites, Nathan. "Political Negotiation as a Process of Modifying Utilities." *Journal of Conflict Resolution* 6 (1962):19–28.

Issacs, Rufas. *Differential Games: A Mathematical Theory with Applications to Warfare and Pursuit, Control and Optimization.* New York: John Wiley and Sons, 1965.

Kahneman, Daniel, and Tversky, Amos. "The Psychology of Preferences." *Scientific American* (January 1982):160–173.

Kant, Immanuel. *Foundations of the Metaphysics of Morals (and "What Is Enlightenment"),* translated by Lewis White Beck. New York: Bobbs-Merrill, 1959. (Originally published 1785.)

Kalisch, G. K., et al. "Some Experimental Games." In *Decision Processes,* edited by R. M. Thrall, C. H. Coombs, and R. L. Davis, pp. 301–327. New York: John Wiley and Sons, 1954.

Kaplan, Morton A. "The Calculus of Nuclear Deterrence." *World Politics* 11 (1958–1959):20–43.

Karlin, Samuel. *Mathematical Methods and Theory in Games, Programming, and Economics,* vols. I, II. Reading, Mass.: Addison Wesley, 1959.

Kaufman, Herbert, and Becker, Gordon M. "The Empirical Determination of Game-Theoretical Strategies." *Journal of Experimental Psychology* 61 (1961):462–468.

Kelley, H. H., and Arrowood, A. J. "Coalitions in the Triad: Critique and Experiment." *Sociometry* 23 (1960):217–230.

Keynes, John M. *Monetary Reform.* New York: St. Martin's Press, 1972.

Kuhn, H. W. "A Simplified Two-Person Poker." In *Contributions to the Theory of Games,* edited by H. W. Kuhn and A. W. Tucker, pp. 97–103. Princeton: Princeton University Press, 1950.

Lacey, Oliver L., and Pate, James L. "An Empirical Study of Game Theory." *Psychological Reports* 7 (1960):527–530.

Lave, Lester B. "An Empirical Description of the Prisoner's Dilemma Game." Rand Memorandum P–2091, 1960.

Lee, M.; McKelvy, R. D.; and Rosenthal, H. "Game Theory and the French Apparentments of 1951." *International Journal of Game Theory* 8 (1979):27–53.

Leiserson, M. A. "Factors and Coalitions in One Part of Japan: An Interpretation Based on the Theory of Games." *American Political Science Review* 62 (1968):770–787.

Lieberman, Bernhardt. "Human Behavior in a Strictly Determined 3x3 Matrix Game." *Behavioral Science* 4 (1960):317–322.

Littlechild, S. C., and Owen, Guillermo. "A Simple Expression for the Shapley Value in a Special Case." *Management Science* 20 (1973):370–372.

Loomis, James L. "Communication, the Development of Trust, and Cooperative Behavior." *Human Relations* 12 (1959):305–315.

Lucas, William F. "A Game with No Solution." *Bulletin of the American Mathematical Society* 74 (1968):237–239.

———. "An Overview of the Mathematical Theory of Games." *Management Science* 18 (1972):3–19.

———, and Thrall, R. M. "N-Person Games in Partition Function Form." *Naval Logistics Research Quarterly* 10 (1963):281–298.

Luce, Duncan R., and Raiffa, Howard. *Games and Decisions.* New York: John Wiley and Sons, 1957.

Luce, Duncan R., and Rogow, Arnold. "A Game Theoretic Analysis of Congressional Power Distribution for a Stable Two-Party System." *Behavorial Science* 1 (1956):83–95.

Lutzker, Daniel R. "Internationalism as a Predictor of Cooperative Behavior." *Journal of Conflict Resolution* 4 (1960):426–430.

———. "Sex Role, Cooperation and Competition in a Two-Person, Non-Zero-Sum Game." *Journal of Conflict Resolution* 5 (1961):366–368.

McClintlock, Charles C., et al. "Internationalism-Isolationism, Strategy of the Other Player, and Two-Person Game Behavior." *Journal of Abnormal and Social Psychology* 67 (1963):631–636.

McDonald, John. "How the Man at the Top Avoids Crises: Excerpts from 'The Game of Business.'" *Fortune* 81 (1970):120.

McGlothlin, William H. "Stability of Choices among Certain Alternatives." *American Journal of Psychology* 69 (1956):604–615.

Mack, David; Auburn, Paulan; and Knight, George P. "Sex Role Identification and Behavior in a Reiterated Prisoner's Dilemma Game." *Psychonomic Science* 24 (1971):59–61.

Markowitz, Harry. "The Utility of Wealth." In *Mathematical Models of Human Behavior,* symposium edited by Jack W. Dunlap, pp. 54–62. Stanford, Conn.: Dunlap Assoc., 1955.

Marshall, James Garth. "Majorities and Minorities: Their Relative Rights." 1853 (pamphlet).

Maschler, Michael. "A Price Leadership Solution to the Inspection Procedure in a Non-Constant-Sum Model of a Test-Ban Treaty." In the Final Report to the U.S. Arms Control and Disarmament Agency under Contract No. ACDA ST–3. Princeton, N. J.: Mathematica Inc., 1963.

May, Mark A., and Doob, Leonard W. "Competition and Cooperation." Social Science Research Council Bulletin No. 25 (1937).

Meer, H. C. van der. "Decision-Making: The Influence of Probability Preference, Variance Preference and Expected Value on Strategy in Gambling." *Acta Psychologica* 21 (1963):231–259.

Miller, Robert B. "Insurance Contracts as Two-Person Games." *Management Science* 18 (1972):444.

Mills, Harlan D. "A Study in Promotional Competition." In *Mathematical Models and Methods in Marketing,* edited by Frank M. Bass, pp. 271–301. Homewood, Ill.: Richard D. Irwin, Inc., 1961.

Miyasawa, K. "The N-Person Bargaining Game." *Econometric Research,* Group Research Memorandum No. 25 (1961).

Moglower, Sidney. "A Game Theory Model for Agricultural Crop Selection." *Econometrica* 30 (1962):253–266.

Morin, Robert E. "Strategies in Games with Saddle Points." *Psychological Reports* 7 (1960):479–485.

Mosteller, Frederick, and Nogee, Philip. "An Experimental Measure of Utility." *Journal of Political Economy* 59 (1951):371–404.

Munson, Robert F. "Decision-Making in an Actual Gaming Situation." *American Journal of Psychology* 75 (1962):640–643.

Mycelski, Jan. "Continuous Games with Perfect Information." In *Advances in Game Theory,* edited by H. W. Kuhn and A. W. Tucker, pp. 103–112. Princeton: Princeton University Press, 1964.

Nash, John F. "The Bargaining Problem." *Econometrica* 18 (1950):155–162.

––––––– . "Two-Person, Cooperative Games." *Econometrica* 21 (1953):128–140.

Neumann, John von, and Morgenstern, Oskar. *The Theory of Games and Economic Behavior,* 3rd ed. Princeton: Princeton University Press, 1953.

Newman, Donald J. "A Model for Real Poker." *Operations Research* 7 (1959):557–560.

Newman, R. G. "Game Theory Approach to Competitive Bidding." *Journal of Purchasing* 8 (1972):50–57.

Niemi, P., and Riker, William H. "The Stability of Coalitions on Roll Calls in the House of Representatives." *American Political Science Review* 56 (1962):58–65.

O'Connor, Richard. *Gould's Millions.* New York: Ace Books, 1962.

Parkison, P. W. "Investment Decision-Making: Conventional Methods vs. Game Theory." *Management Accounting* 53 (1971):13–15.

Parthasarathy, T., and Raghavan, T. E. S. "Some Topics in Two-Person Game Theory." *Modern Analytic and Computational Methods in Mathematics.* No. 22, New York: American Elsevier Co., 1971.

Poe, Edgar Allan. *The Works of Edgar Allan Poe,* vol. 2. New York: Harper & Bros., 1902.

Polsby, Nelson W., and Wildavsky, Aaron B. "Uncertainty and Decision-Making at the National Conventions." In *Political and Social Life,* edited by N. W. Polsby, R. A. Dentler, and P. A. Smith, pp. 370–389. Boston: Houghton Mifflin Co., 1963.

——— . *Presidential Elections: Strategies of American Electoral Politics.* New York: Charles Scribner's Sons, 1964.

Preston, Malcolm G., and Baratta, Phillip. "An Experimental Study of the Auction-Value of an Uncertain Outcome." *American Journal of Psychology* 71 (1958):183–193.

Raiffa, Howard. "Arbitration Schemes for Generalized Two-Person Games." Report M720-1 R-30 of The Engineering Research Institute, University of Michigan, 1951.

Rao, Ambar G., and Shakun, Melvin F. "A Quasi-Game Theory Approach to Pricing." *Management Science* 18 (1972):110–123.

Rapoport, Anatol. *Fights, Games and Debates.* Ann Arbor: University of Michigan Press, 1960.

——— , and Orwant, Carol. "Experimental Games: A Review." *Behavioral Science* 7 (1962):1–37.

Rapoport, A.; Guyer, M.; and Gordon, D. "A Comparison of Performance of Danish and American Students in a 'Threat Game.'" *Behavorial Science* 16 (1971):456–466.

Read, Thornton. "Nuclear Tactics for Defending a Border." *World Politics* 15 (1963):390–402.

Richter, Marcel K. "Coalitions, Core and Competition." *Journal of Economic Theory* 3 (1971):323–334.

Riker, William H. *The Theory of Political Coalitions.* New Haven: Yale University Press, 1962.

——— . "A Test of the Adequacy of the Power Index." *Behavorial Science* 4 (1959):120–131.

Robinson, Frank D. "The Advertising Budget." In *Readings on Marketing,* edited by S. George Walters, Max D. Snider, and Morris L. Sweet. Cincinnati: Southwestern Publishing, 1962.

Robinson, Julia. "An Iterative Method of Solving a Game." *Annals of Mathematics* 54 (1951):296-301.

Rose, Arnold M. "A Study of Irrational Judgments." *The Journal of Political Economy* 65 (1957):394-402.

Rosenthal, Robert W. "External Economies and Cores." *Journal of Economic Theory* 3 (1971):182-188.

Royden, Halsey I.; Suppes, Patrick; and Walsh, Karol. "A Model for the Experimental Measurement of the Utility of Gambling." *Behavioral Science* 1 (1959):11-18.

Sankoff, D., and Mellos, S. "The Swing Ratio and Game Theory." *American Political Science Review* 66 (1972):551-554.

Savage, Leonard J. *The Foundations of Statistics.* New York: John Wiley and Sons, 1954.

Scarf, Herbert E., "On the Existence of a Cooperative Solution for a General Class of n-Person Games." *Journal of Economic Theory* 3 (1971):169-181.

Schelling, Thomas C. "The Strategy of Conflict—Prospectus for a Reorientation of Game Theory." *Journal of Conflict Resolution* 2 (1958):203-264.

―――. "Bargaining, Communication and Limited War." *Journal of Conflict Resolution* 1 (1957):19-36.

Schubert, Glendon A. *Quantitative Analysis for Judicial Behavior.* Glencoe, Ill.: Free Press, 1952.

―――. *Constitutional Politics.* New York: Holt, Rinehart and Winston, 1960.

Schwartz, E., and Greenleaf, J. A. "A Comment on Investment Decisions, Repetitive Games, and the Unequal Distribution of Wealth." *Journal of Finance* 3 (1978):122-127.

Scodel, Alvin. "Induced Collaboration in Some Non-Zero-Sum Games." *Journal of Conflict Resolution* 6 (1962):335-340.

―――. "Probabilty Preferences and Expected Values." *Journal of Psychology* 56 (1963):429-434.

―――, and Minas, J. Sayer. "The Behavior of Prisoners in a 'Prisoner's Dilemma' Game." *Journal of Psychology* 50 (1960):133-138.

―――, and Ratoosh, Philburn. "Some Personality Correlates of Decision Making under Conditions of Risk." *Behavioral Science* 4 (1959):19-28.

Scodel, Alvin, et al. "Some Descriptive Aspects of Two-Person, Non-Zero-Sum Games." *Journal of Conflict Resolution* 3 (1959):114-119.

Bibliography

Scodel, Alvin, et al. "Some Descriptive Aspects of Two-Person, Non-Zero-Sum Games, II." *Journal of Conflict Resolution* 4 (1960):193–197.

Shapley, L. S. "A Value for n-Person Games." In *Contributions to the Theory of Games*, edited by H. W. Kuhn and A. W. Tucker, pp. 307–317. Princeton: Princeton University Press, 1953.

——— , and Shubik, Martin. "A Method for Evaluating the Distribution of Power in a Committee System." *American Political Science Review* 48 (1954):787–792.

——— . "On the Core of Economic Systems with Externalities," *American Economic Review* 59 (1969):678–684.

Shubik, Martin. *Strategy and Market Structure.* New York: John Wiley and Sons, 1960.

——— . "The Dollar Auction Game: A Paradox in Noncooperative Behavior and Escalation." *Journal of Conflict Resolution* 15 (1971):109–111.

Siegel, Sidney, and Harnett, D. L. "Bargaining Behavior: A Comparison Between Mature Industrial Personnel and College Students." *Operations Research* 12 (1964):334–343.

Simmel, Georg. *Conflict and the Web of Group Affiliations*, translated by Kurt H. Wolff and Reinhard Bendix. Glencoe, Ill.: Free Press, 1955. (Originally published 1908.)

Simon. H. A. "Theories in Decision-Making in Economics and Behavioral Science." *American Economic Review* 49 (1959):253–283.

Smith, J. M. "The Evolution of Behavior." *Scientific American* 239 (1978):176–192.

——— , and G. A. Parker. "The Logic of Assymetric Contests," *Animal Behavior* 24 (1976):159–175.

Smithies, A. "Optimum Location in Spatial Competition." *Journal of Political Economy* 49 (1941):423–429.

Snyder, Glenn H. "Deterrence and Power." *Journal of Conflict Resolution* 4 (1960):163–178.

Sommer, R. "The District Attorney's Dilemma: Experimental Games and the Real World of Plea Bargaining." *American Psychologist* 37 (1982):526–532.

Stone, Jeremy. "An Experiment in Bargaining Games." *Econometrica* 26 (1958):282–296.

Straffin, Phillip D., Jr. "The Bandwagon Curve." *American Journal of Political Science* 21 (1977):695–709.

——— , and Heaney, J. P. "Game Theory and the Tennessee Valley Authority." *International Journal of Game Theory* 10 (1981):35–43.

Straffin, P. D., Jr.; Brams, S. J.; and Davis, M. D. "Power and Satisfaction in an Ideologically Divided Body." In *Power, Voting and Vot-*

ing Power, edited by M. J. Holler, pp. 239–255. Wurzburg: Physics-Verlag, 1981.

Stryker, S., and Psathas, G. "Research on Coalitions in the Triad: Findings, Problems and Strategy." *Sociometry* 23 (1960):217–230.

Suppes, Patrick, and Walsh, Karol. "A Non-Linear Model for the Experimental Measure of Utility." *Behavorial Science* 4 (1959):204–211.

Thompson, G. L. "Bridge and Signaling." In *Contributions to the Theory of Games,* edited by H. W. Kuhn and A. W. Tucker, pp. 279–290. Princeton: Princeton University Press, 1953.

Thorpe, Edwin. *Beat the Dealer.* New York: Vintage Books, 1966.

———, and Waldman, W. E.. "The Fundamental Theorem of Card Counting with Applications to Trente-et-Quarante and Baccarat." *International Journal of Game Theory* 2 (1973):109–119.

Tropper, Richard. "The Consequences of Investment in the Process of Conflict." *Journal of Conflict Resolution* 16 (1972):97–98.

Uesugi, Thomas, J., and Vinacke, Edgar W. "Strategy in a Feminine Game." *Sociometry* 26 (1963):75–88.

Uslaner, Eric M., and Davis, J. Ronnie. "The Paradox of Vote-Trading: Effects of Decision Rules and Voting Strategies on Extremalities." *American Political Science Review* 69 (1975):929–924.

Vickrey, William. "Self-policing Properties of Certain Imputation Sets." In *Contributions to the Theory of Games,* edited by A. W. Tucker and R. D. Luce, pp. 213–246. Princeton: Princeton University Press, 1959.

Vinacke, Edgar W. "Sex Roles in a Three-Person Game." *Sociometry* 22 (1959):343–360.

Weinberg, Robert S. "An Analytic Approach to Advertising Expenditure Strategy." In *Mathematical Models and Methods in Marketing,* edited by Frank N. Bass et al., pp. 3–34. Homewood, Ill.: Richard D. Irwin, (1961). (Originally published by the Association of National Advertisers, 1960.)

Willis, Richard H., and Joseph, Myron L. "Bargaining Behavior I: 'Prominence' as a Predictor of the Outcomes of Games of Agreement." *Journal of Conflict Resolution* 3 (1959):102–113.

Wilson, Robert, "Stable Coalition Proposals in Majority-Rule Voting." *Journal of Economic Theory* 3 (1971):254–271.

Index

Index